T0301584

Redeem All

Redeem All

HOW DIGITAL LIFE IS CHANGING
EVANGELICAL CULTURE

Corrina Laughlin

UNIVERSITY OF CALIFORNIA PRESS

University of California Press
Oakland, California

© 2022 by Corrina Laughlin

Cataloging-in-Publication Data

Names: Laughlin, Corrina, 1984– author.
Title: Redeem all : how digital life is changing evangelical culture /
    Corrina Laughlin.
Identifiers: LCCN 2021022883 (print) | LCCN 2021022884 (ebook) |
    ISBN 9780520379671 (cloth) | ISBN 9780520379688 (paperback) |
    ISBN 9780520976856 (epub)
Subjects: LCSH: Digital media—California—Santa Clara Valley (Santa
    Clara County)—Religious aspects—Christianity. | Information technol-
    ogy—California—Santa Clara Valley (Santa Clara County)—Religious
    aspects—Christianity. | Church and mass media—California—Santa
    Clara Valley (Santa Clara County) | Evangelicalism—California—Santa
    Clara Valley (Santa Clara County) | BISAC: BUSINESS &
    ECONOMICS / Nonprofit Organizations & Charities / Marketing &
    Communications | TECHNOLOGY & ENGINEERING / Mobile &
    Wireless Communications
Classification: LCC BL265.I54 L38 2021 (print) | LCC BL265.I54 (ebook)
    | DDC 261.5/2—dc23
LC record available at https://lccn.loc.gov/2021022883
LC ebook record available at https://lccn.loc.gov/2021022884

30 29 28 27 26 25 24 23 22 21
10 9 8 7 6 5 4 3 2 1

Contents

Acknowledgments

First and foremost, this book could not have come about without the generosity and openness of the people quoted throughout. Over a decade-long period, Christians all over the country invited me into their homes, offices, and churches—and for that I am incredibly grateful.

This project started as a master's thesis at New York University, and without the support of Thomas Augst it would not have even begun. It continued at the Annenberg School for Communication at the University of Pennsylvania under the incomparable guidance of John L. Jackson Jr. Along with John, my dissertation committee at Annenberg—Carolyn Marvin, Guobin Yang, and Anthea Butler—provided invaluable input and introduced me to methodological and theoretical frameworks that guided my thinking. Throughout my time at Annenberg, I was privileged to be exposed to the teaching and research of exemplary scholars, including Sharonna Pearl and Barbie Zelizer. My colleagues provided and continue to provide inspiration. I am especially indebted to Sun-ha Hong and Aaron Shapiro, who helped me write the proposal for this book, and to Lee McGuigan, who offered support and advice as I was writing it.

I have also been supported by incredible, inspiring colleagues at Loyola Marymount. I especially appreciate Michele Hammers and Meng

Li, who spent time reading and offering feedback on parts of the manuscript. Outside of my home university, a rotating cast of stellar scholars whom I consider my religion and media "tribe" helped me conceive of this book and have provided great feedback and advice. In particular, I thank Stewart Hoover, who has been a kind and caring mentor.

Everyone at the University of California Press has been lovely to work with, especially editors Lyn Uhl and Michelle Lipinski, who guided me through this process. I would not and could not have finished this book without the support of Gaby Gil, Rocio, and Manny, who kept their daycare open during the COVID-19 pandemic and who became our central support system during a traumatic year. My parents, Tim and Teresa Laughlin, and my in-laws, Tom and Sharon Merchant, have also provided support and care.

Last, I thank my husband, Brian Merchant, who has remained my best booster and editor, and my beautiful children, Aldus and Russell.

Introduction

Christians should be at the head of innovation and not the
tail. We shouldn't wait for Google or for Apple to come out
with Christian innovation, it should be the Christians
trying to innovate for what is needed in the Kingdom.

—Jeyanti Yorke

The day that the iPhone came out I was in Las Vegas at the
Consumer Electronics Show in the Nokia tent. Heard the
announcement... and I hit publish on a page that said how
to get your Bible onto an iPhone.

—Antoine Wright

In the summer of 2013, in an underground conference room on the base-
ment level of a labyrinthine megachurch in Dallas, Texas, I watched a
Christian social media consultant give a presentation to a room full of
evangelicals. During his PowerPoint talk, which focused on how churches
could incorporate social media into their outreach strategies, he told the
audience: "We've never been more equipped and more resourced to get
the message of the gospel out there.... I really do believe that we could be
part of that generation or part of raising up the next generation that could
see Christ return. And we have an amazing opportunity, but we're going to
be held accountable for how we stewarded what God gave to us."

In the cosmic play of Christian history, many evangelicals see the contemporary moment as the final or penultimate act before the return of
Christ, or the Rapture. As such, Christians living on Earth during this
time believe that they have a responsibility to do everything in their power

to fulfill what they call "the Great Commission," the biblical imperative to bring the gospel, or the message of the Bible, to all people. As this social media consultant suggested to his audience, individual believers will be judged by God based on how effectively they can harness these God-given tools to proselytize and convert nonbelievers through religious apps, VR experiences, Instagram stories, podcasts, and beyond. This book, *Redeem All*, tells the stories of those passionate American evangelical media makers whose work is inspired by this directive. These Christians want to redeem the internet, to redeem Silicon Valley, to redeem evangelical culture, to redeem the globe, and in every instance their work lines up with spiritual principles and purposes. To these ends, American evangelicals have innovated, hacked, lauded, and adapted digital media technologies, but I argue that as they embrace what I call digital habitus, they have opened Pandora's box, releasing new authorities, forms, and discourses that have changed evangelical culture forever.

This book represents a decade of qualitative research guided by the ethnographic principle to "follow the habitus"—a remix of the anthropologist George Marcus's (1996) famous dictum. Marcus wrote at a time when increasing globalization rendered the ethnographic ideal of a "field site" obsolete. For Marcus, ethnographers needed to adapt by following their subjects across borders in real and virtual ways. This research principle has led me to conferences in megachurches, to online chatrooms for millennial evangelicals, to Twitter hashtags, to home offices of prominent Christian technologists, to coffee shops, to Silicon Valley start-ups, to the cafeterias of big tech companies where religious tech workers toil, to churches in the American South pioneering online community platforms, to towns in New Jersey, to my southern California hometown, to Oklahoma, to New York City.

The decade during which I undertook this project was punctuated by events that led me to reimagine the story I was writing. I witnessed the shifting evangelical response to the legalization of same-sex marriage— what had originally seemed to be a central wedge issue that fueled the culture wars was sidelined. New Christian celebrities arose on Instagram, YouTube, and on podcasts, reorienting the evangelical media landscape. The 2016 election of Donald Trump upended the widespread assumption about evangelicals drifting to the political center. The #MeToo movement sparked passionate conversations about sexual abuse and harass-ment in evangelical culture. The COVID-19 crisis saw evangelicals further embracing online church and other digital tools. And the 2020 uprisings against police brutality created space for another jolt in the conversation around race within the evangelical church. At each turn, I watched as these moments played out dramatically over social media outlets such as Twitter, Facebook, and Instagram. I talked through new ideas with pas-sionate Christian media makers. And I watched American evangelical cul-ture change, post by post.

WHAT IS THE EVANGELICAL?

There's no clear consensus about what, if anything, evangelicalism is. Although scholars and preachers alike have tried to define the term, it eludes simple descriptions. The media often focuses on the most sensa-tional and salacious aspects of evangelical culture, and as such, this is how much of the world imagines American evangelicals. Some people think first of the prominent televangelists who were plagued by scandals in the 1980s—the term conjures the image of Tammy Faye Bakker's mascara-stained face. Others think of the Robertson family of raucous, religious hunters portrayed in the reality show *Duck Dynasty*. And while I was writing this book, I had at least one person ask how Westboro Baptist Church—the organization known for picketing military funerals holding signs and yelling inflammatory statements fit into my study.[1] Westboro is not an evangelical organization—in fact, many people would categorize it as a new religious movement, or a cult—but this misunderstanding shows how secular culture often views evangelicals through the lens of the media and in wildly different ways than evangelicals view themselves.

In my journeys in evangelical culture I have met earnest believers who upend the typical assumptions and media tropes that have come to define evangelicals in the popular American imagination. I have learned that many of the mainstream narratives about evangelicals obscure the com-plex realities of this religious subculture. This book is not an apologetic text. I make no claims as to whether evangelicals are good or bad—like any religious subculture they are complicated, and they are certainly more complicated than media tropes about them allow for. Although evangelicals are often portrayed in the media as fundamentalist Protestants who harbor a strong, politically conservative bent, there are many strands of evangelical Christianity—from the conservative soccer (or hockey) moms perhaps most famously embodied by Sarah Palin to televangelist preachers promoting the so-called "prosperity gospel" (see Bowler, 2013; Walton, 2009), to rave-throwing millennial evangelicals in the Hamptons (see Kisner, 2013).

Though many people connect evangelicalism to a Southern, parochial worldview, the beginnings of modern evangelicalism could just as easily be traced to Los Angeles, to the historical Azusa Street Revival or to the hard-to-categorize "Jesus Freak" scene. And we would not be telling the full story of evangelical history if we excluded the traditions of the Black church. Nor can we bound evangelicalism at the U.S. border. In fact, one characteristic of evangelicalism in America might be that it does not have a solid ground on which it pitches its wide tent. Some scholars have even argued that because of its diversity of forms, evangelicalism qua evangelicalism does not exist.[2]

Although evangelicalism comes in many shapes, when Americans call themselves "evangelical," what they often mean is that they have an affective, emotional relationship with God and Jesus, as the anthropologist Tanya Luhrmann (2012) found in her study of evangelical worship. This typically comes in the form of an emphasis on the Bible as the literal Word of God, a belief in the power of prayer, a drive to spread their faith by "witnessing" to others, and an adherence to following moral and spiritual dictums of Jesus—popularly and succinctly explained in the acronym WWJD: What Would Jesus Do?[3] Beyond this, evangelicalism in America is an assemblage of cultural and theological norms that are recognizable to anyone who has spent a few Sundays in their local megachurch. Evangelical culture can also be characterized with reference to its robust media footprint—the popular music of the Australian megachurch Hillsong played in churches and Christians spaces; the celebrities it claims, such as Justin Bieber and Chris Pratt; the books, bumper stickers, and apparel that dot the American pop culture landscape.

Evangelical culture is everywhere in America and it does not hide, although it sometimes purposefully cloaks itself in popular cultural forms.

Because the First Amendment's Establishment Clause forbids national-ized religion, religious entities in America have appealed to the populace rather than to the state. As many scholars have noted, there has been no more successful populist religion in America than evangelicalism.[4] Whether by preaching outdoors, on the radio, or on television, American evangelicalism has historically been invested in understanding, keeping up with, and often mimicking popular culture, and deploying, adapting, and appropriating the technologies that transmit it.

At the same time, evangelicals hope to be countercultural. They talk about the imperative to be "in but not of the world." By this they mean that they must understand and participate in worldly things so as to be able to connect to the "unchurched," but as individuals and organizations, evan-gelicals must not embrace what they see as the sinful nature of secular culture. Because of this cultural bent, evangelicalism reflects popular cul-ture. This aspect of evangelicalism has been explored by scholars working at the intersection of media studies and religious studies who have rightly pointed out that evangelicals have been early adopters of media tech-nologies and, in many cases, have advanced and helped to define media forms.[5] And Heather Hendershot (2004) has described how evangelicals have created a vast, complex, and lucrative media and material culture. This book follows in the tradition of media scholars who have focused on evangelicals by exploring the evangelical understanding and embrace of new media technologies.

As such, the story this book tells is not the *political* story of evangeli-calism, although it does intersect with and draw from that story. From its roots in the Scopes Monkey Trial, to the anticommunist rhetoric and pro-business theology of Billy Graham, to the rise of California Republicans and Ronald Reagan, to the seemingly solid support of white evangelicals for Donald Trump, the twists and turns of evangelical politics have been well mapped by others.[6] Instead, the story this book tells is about evangel-ical *culture*. Of course, culture can never be fully separated from politics. Evangelical cultural and media products have been the means through which evangelicals define themselves and maintain the sometimes porous boundary between their communities and the secular world.[7] So culture is often a site for contestation, and as evangelicals approach new media technologies and establish new institutions, ideas, and forms, they perform this contestation. As they establish new spaces, and new practices in digital culture, I argue, evangelicals push, warp and change the boundaries of their subculture.

MEDIA, TECHNOLOGY, AND EVANGELICAL CULTURE

A persistent myth about evangelicalism arose during the Scopes Trial in 1925. Evangelicals, the myth goes, are antiscience, antitechnology, antiprogress. As the famous H. L. Mencken of *The Baltimore Sun* wrote about fundamentalist Christians: "Every valuable thing that has been added to the store of man's possessions has been derided by them when it was new, and destroyed by them when they had the power. They have fought every new truth ever heard of, and they have killed every truth-seeker who got into their hands" (1925). Mencken's commentary came at a time when the so-called Monkey Trial saw the antievolution cause fought and won by the famous litigator William Jennings Bryan. Although Bryan's arguments carried the day and he won the right to ban teaching evolution in the Tennessee curriculum, fundamentalists lost the battle of public opinion. Mencken's scathing characterization of fundamentalists as "backward" stuck and became the commonly understood connotation of fundamentalism. Later, despite many attempts to shake off this antiprogress reputation, it remained a central aspect of the public perception of evangelical culture.

As fundamentalists evolved and rebranded in the 1940s, they self-consciously used the new medium of radio to try to buck against this stereotype of conservative Christians as antimodern. Radio helped fundamentalist Protestantism gain a new national identity and furthermore, through their participation on the airwaves, Christians could claim that they too had a place in contemporary American culture. Although firebrand preachers like the anticommunist crusader Carl McIntyre gained national popularity, many Christians self-consciously defined themselves against the rabid fundamentalism McIntyre embodied and began to call themselves "neo-evangelicals." Neo-evangelicals had learned from the lessons of the Scopes Trial, chief among them to avoid using the term "fundamentalist."[8] Much like the later moniker "compassionate conservative," "neo-evangelical" denoted a kinder, gentler Christianity. These evangelicals still believed in the tenets of fundamentalism, especially in the inerrancy of the Bible, but they softened their tone and attempted to become more inclusive. They wanted to shift their public image from the stone-faced guardians of literalism and tradition to that of a cadre of friendly folks spreading the Good News. In 1942 this new brand of evangelicalism gained a public face, with the founding of the National Association of Evangelicals (NAE). And at this time, the inchoate movement was defined in the public sphere primarily by new voices transmitted through radio waves—a medium that helped spread the message but also represented the modernity of which evangelicals saw themselves a part.[9]

Also in the 1940s, Billy Graham, a young preacher from North Carolina who cut his teeth at the parachurch organization Youth for Christ, began to gain national notoriety. Staunch fundamentalists criticized what they perceived as Graham's openness, but the image that Graham embodied of the wholesome, clean-cut, all-American Christian appealed to Cold War Americans en masse. Graham would exert an influence on national politics and evangelical culture until the early years of the twenty-first century. One of his strengths was that he understood that media technologies were crucial tools for solidifying evangelical identity. In the mid-twentieth century Graham set up a network of Christian media outlets. In 1956, with his father-in-law, he founded the evangelical magazine *Christianity Today*. In the same spirit Graham threw his name behind myriad business ventures, including his film production house Billy Graham Films (later World Wide Pictures) and the popular radio program *The Hour of Decision*.

Because this strategy proved to be successful, evangelicals began to open up new cultural spaces for themselves in which they could be assured that their values would be respected. For example, James Dobson founded Focus on the Family in 1977, an organization devoted to promoting "family values." Focus on the Family became a multi-million-dollar business by selling Christian-themed media products, such as magazine subscriptions, to an evangelical audience. Their success proved that catering to an evangelical audience who sought an alternative media culture was a savvy business decision. Thus, evolving understandings of the evangelical audience allowed Christians to identify as evangelicals not only on the local scale through church involvement but also through the habitus engendered by consumerism and media engagement. This trend continued in the 1980s with televangelism and in the 1990s with the boom in Contemporary Christian Music (CCM), a genre that tweaked popular musical styles and added Christian messages as a way to remain relevant to a young Christian audience.

Continuing their historical engagement with technology, as the internet and social media became drivers of American cultural engagement in the early aughts, evangelicals were there. As Heidi Campbell (2010) has noted in her study of how religious organizations approach new media, evangelicals were more receptive to technology than other religious traditions because they felt that "the goal of evangelism that can be realized through this technology seems in many respects to outweigh the criticism and cautions raised" (p. 39). As a tradition that has enthusiastically embraced mass media technologies in the past, this is no surprise, and indeed is consistent with the history of evangelical engagement in the public sphere.[10] The consensus within evangelical culture has been that although technology can be dangerous and corrupting, evangelicals must use it and shape it to their ends, as they have throughout their history. They have historically used popular media technologies to attract spiritual seekers and to prove that their brand of Christianity can keep up with changing fashions in the secular world. As in the 1940s with radio, the 1970s with publishing, the 1980s with television, and the 1990s with music, today's evangelicals want to use technology as a means to prove that their message still has a viable place in the modern world. In an age when digital habitus pervades American culture, it is no surprise that evangelical Christians would be on the forefront of developing technologies for Christian audiences.

DIGITAL HABITUS

Redeem All charts how the widespread adoption and integration of digital technologies in churches, organizations, and in the lives of believers is affecting evangelical culture. I argue that the resultant habitus is reshaping what it means to be an evangelical. For the sociologist Pierre Bourdieu (1977), the term "habitus" encompasses all of the socially and culturally

conditioned practices that define daily life, from the way a person uncon-
sciously picks up a fork to the frequency with which she checks the Twitter
app on her iPhone. Our behaviors are learned from childhood through
the experiences that teach us to understand our social world. "Habitus"
is the way that a culture or society replicates through individual and collective practices, but it is not fixed or externally imposed and may change
over time.[11] We are all raised into cultures with values that we come to
understand as normal, and our habits and behaviors enforce these values
to ourselves and others.

Bourdieu tells us that habitus works on us in unconscious ways; we
think of the thoughts and actions that define our habitus as common
sense when we think of them at all. But Bourdieu explains that habitus is
situated within class structures and helps to perpetuate class boundaries
because, for Bourdieu, habitus is a disciplining force. What may be common sense to a working-class person is not necessarily common sense to an
elite. As Bourdieu wrote: "The class *habitus* is nothing but this *experience*
(in its most usual sense) which immediately reveals a hope or an ambi-
tion as reasonable or unreasonable, a particular commodity as accessible
or inaccessible, a particular action as suitable or unsuitable" (1965/1990,
p. 5, emphasis in the original). Habitus creates harmony inside of classes
and discord when class boundaries are transgressed through social pressures that are well understood by in-group members despite being rarely
explicit or even conscious.

Technologies too are socially and historically shaped. They emerge
from cultures and live in social systems, and as such, they both guide
habitus and *are* habitus. As Jonathan Sterne put it, "understood socially,
technologies are little crystallized parts of habitus" (2003, p. 376). As digital, mobile media becomes more ubiquitous, what I term "digital habitus"
has infused American culture. All of the daily interactions and micro-
interactions that are facilitated by digital media can be characterized as
digital habitus. This habitus is often guided by digital platforms such as
Facebook, Twitter, or Instagram, which encourage repetitive social behavior such as scrolling, liking, posting, and engaging. For example, when my
oldest child was three years old, he started referring to a pretend smartphone on his hand, checking his "map" whenever he wanted to confirm
or refute something. "Pick up your toys," I would say and he would look

at his palm and tell me, "No, Mommy. My map says it is not time to do that yet." He had learned, without being formally taught, that having a smartphone in your hand allowed a person to make decisions or place themselves in space. His understanding of the "map" was the result of his constant interaction with adults preoccupied by their phones. This small example reveals how technologies become cultural forms integrated into how we understand what it means to be an actor in the social world. Of course, my three-year-old was also revealing a class-based understanding of technology. He was emulating the habits of his middle-class American parents.

And indeed digital habitus reveals and is revealed by a classed understanding of technology and technological progress. Habitus is generative, it creates culture. Digital habitus creates digital culture; it reproduces offline social worlds in the somewhat distorted online world. In this translation, habitus creates new behaviors, norms, and modes of being in the world. New forms, like memes circulating on Instagram, emojis with multiple meanings, jokes, mores, hierarchies, and vocabularies are understood by some groups and remain inscrutable to others. This is generational—there is a whole cohort of people we call "digital natives," a term meant to signal the way that younger people seem to naturally understand the digital social world. Digital habitus is also class-based. Educational leaders focus on the digital divide—the fear that lower-income people have less digital fluency and less access to digital tools than higher-income people, or the corollary fear that low-income children are getting too much "screen time" at the expense of their education.

Apart from unequal access to digital tools, class and digital habitus are inextricably linked in terms of aesthetic principles. This doesn't mean that only certain classes of people use computers and smartphones, just that the rules of the production and distribution, display, and exchange of these tools and what they produce are different among social groups, even though enthusiastic technologists often claim that their products are culturally neutral, or universally pleasing. For example, there is a common perception that Apple products are particularly "user friendly."

One need not be an early adopter of technology or an enthusiastic techie to be entrenched in digital habitus. In her history of internet culture, Joanne McNeil (2020) has made the case that in trying to produce a "mirror of the world" (a phrase she quotes from former Google VP Marissa Mayer), Google and other technology companies have *altered* the world, creating a new twenty-first-century subjectivity—that is, those who were once people, McNeil asserts, are now "users." We now live in the mirror world, whether we like it or not. Our homes have been photographed and mapped, our knowledge, ideas, and photographs have been stored and indexed, even our voices have been logged and used to train thinking machines. We are defined by the condition of being a "user" whether we enthusiastically or uneasily adopt the identity of one.

When we master digital habitus, we thus prove our fitness in American culture. When we fail to use digital tools, cannot use them, or refuse to use them, there are disciplinary responses. We might not be able to get the job we wanted because our netiquette is lacking polish. We might not be able to communicate with our children or grandchildren. We might miss out on gossip, on news, on relationships. We might be casually mocked or dismissed as a luddite when we refuse to keep up with changing consumer media technologies.[12] Scholars of internet history have characterized the birth of the internet and of the later web as one that glorified a particular type of person—a highly educated, white, middle-class man. That it has now become important to have a passport to the worlds these men built means that American social life relies on an exclusionary habitus that values mastery of digital tools over alternate modes of being and communication.

Given all of this, what does it mean when evangelical institutions actively cater to digital habitus as a means of remaining a relevant force in American culture and of proving their own fitness? And furthermore, how has American evangelical culture grappled with the seismic social shifts digital habitus has engendered? This book is centered on these questions.

JOURNEYS IN EVANGELICAL CULTURE:
FOLLOWING THE HABITUS

My research has taken a multimethod approach to understanding the evangelical embrace of digital habitus. I've conducted more than sixty formal interviews with tech entrepreneurs, church leaders, missionaries, writers, and speakers and many more informal interviews with churchgoers and technologists. I've attended evangelical conferences. I worked for a short period of time as an intern at a technology start-up run by evangelicals. I conversed with evangelicals on Twitter and followed influencers on Instagram. I've attended church services all over the country and online. I have immersed myself in the evangelical media sphere, reading books, listening to podcasts, and watching social media livestreams.

Throughout the ten years that I focused on this project, I have been inspired by the guide laid out by the anthropologist George Marcus. Marcus (1996) urged scholars conducting ethnographic research in an increasingly global world to expand their field sites and imagine "following" as a central principle of fieldwork. In a moment when people may be united by ideas, cultural products, and media that cross borders and date lines, Marcus opened up ethnography, once focused on specific, geographically bounded field sites, to include following the people, following the thing, and following the metaphor as ways of connecting dispersed field sites into a multisited ethnographic project.

Because my project focuses on a vast and loosely structured subculture, Marcus's methodology is one that resonates with me. In my research I followed the habitus, concentrating on how and where evangelicals used digital media, in real spaces such as churches as well as online spaces such as social media platforms. Following digital habitus helped me to connect work and discourses promoted by church leaders to creative projects like podcasts produced by community members. This book is about producers and users—those at the top of the evangelical power structure, those at the bottom, and those in-between. This focus on habitus helped me understand and parse the complicated racial politics of American Christianity. With some exceptions, academic works on evangelicalism have often focused on only white evangelicals or only Black evangelicals. Scholars have justified drawing these boundaries along racial lines because they argue that white evangelicalism has been a distinct cultural and political force in the United States with a history that is markedly different from what is known as "the Black church."

Jonathan Walton (2009), however, has explained that this bias in the literature is no longer relevant (p. 20). I take his critique seriously, and I did not set out to study only white Christians. I attended services where popular Black pastors preached to primarily Black audiences, including the inimitable televangelist T. D. Jakes. I sat in on services with Asian and Latino congregations. I attended multiethnic churches as well. Yet I found that the people who shared a zeal for technology were overwhelmingly white evangelicals. This book unpacks the reasons why there are a dearth of Black voices in much of the movement that I trace. I argue that in accepting the discourses of Silicon Valley digital culture, evangelicals have also accepted the biases of this culture, specifically the historical erasure of racial and gender difference in technology production as well as the class-based aesthetics that they create and carry. Both evangelicalism and tech purport to be colorblind spaces that will accept anyone.[13] Yet, as scholars have shown, and as this book argues, this is not the case in either context. This book is about a mind-set that I believe to be particularly American and more specifically endemic to white America. However, digital habitus also cloaks itself in neutrality and normativity. It was rare to hear evangelicals consider racial difference as a factor in their use of technology. In fact, many assume that the more technologically advanced they can be, the more diverse the audience and the reach can and will be.

Evangelicalism has a fraught, contested history of racial exclusion and separation. Although prominent evangelicals like Rick Warren and Russell Moore have urged evangelicals to attempt to heal the racial schisms that have defined the movement in the past, there is little evidence that evangelicalism is becoming more integrated (see Bracey & Moore, 2017). Women, too, face systemic challenges that bar them from occupying leadership roles (chapter 4 dives into this history). It is perhaps because of their own history that evangelicals do not see the faults in the technology industry that are clearly there. And it may be because of this history that white evangelicals in particular are willing to accept the premises of the tech industry more than those evangelicals of color. But even as they are backgrounded in technological production, women and Black Christians have created their own counterpublic spaces in evangelical digital culture. As church leaders and pastors take up ideas coming from Silicon Valley and integrate them into the way that they run their churches, construct their communities, and reach out to the broader world in the hopes of modernizing American evangelical culture, they also empower new voices as they normalize digital habitus.

Digital habitus has proven to be a double-edged sword for evangelical culture. While institutions hope that catering to the digital habitus in American culture will help them remain relevant and grow into the twenty-first century, it has also given more power to voices that were previously siloed. Social media can act as an accelerant for social movements—on Facebook and Twitter ideas can circulate quickly.[14] This is what we think of as the "viral" nature of new media. And given that evangelical culture is a space that has centered on ideas and personalities, it has been rocked by new thought leaders emerging on social media, many of whom would not be allowed to hold the microphone in a conventional church space.

ETHNOGRAPHIC SINCERITY

As John L. Jackson Jr. has noted in *Thin Description* (2013), ethnography no longer has a "backstage," a place an ethnographer can retreat to and keep hidden from his subjects. A researcher can no longer be simply an observer, he is also observed by his informants. This may be particularly true of researchers trying to understand digital contexts and immersing themselves in digital worlds. Unless I were to actively use "dummy accounts" or online alter egos to contact informants (which I never did), my online social life is a reflection of my real life and my ethnographic work. On Twitter and Instagram I follow and converse with friends and family at the same time as I follow and converse with informants. These worlds are not meaningfully separated. I had the experience of introducing myself to an interviewee only to have him allude to details of my life that I had not disclosed to him but had disclosed on social media. This experience and others made me realize that I was being watched and judged.

I am relatively sure that some of my informants have read my degree of "nativeness" in evangelical culture differently than others. In my first year of high school I became a Christian, or "was saved," although this experience, guided by a friend's mother, was not entirely consensual on my part. I had become friends with a group of girls who all went to the same evangelical church. I went to church with them and began to frequent the church's youth group. Our social life revolved around churches for a time. We went to other churches to see Christian bands play or to hear popular preachers' sermons. These girls remained my friends throughout high school and beyond—one of them was the maid of honor at my wedding. As an undergraduate I majored in religious studies and continued to explore my spirituality, although I had stopped going to church on a regular basis several years before and I no longer considered myself a Christian.

When I embarked on this project, it was out of scholarly curiosity, but memories from my church days came back to me. Because of this history I understood when people slipped shorthand versions of Bible verses into their language, and I felt relatively fluent with evangelical jargon. But beyond that, throughout this research, I had genuine experiences in churches that I have considered personally important to my spiritual life: I have openly cried as I watched adults get baptized at a church service in Nashville; I have had deep conversations about faith and life with Christians and pastors; I have rethought many of the beliefs that I had about religion and spirituality. This project opened me up to a world that I had only known as a child and made me consider my faith and spiritual life in a new way—though of course that was not my intent when I began this investigation.

I tried to be as honest as possible when people asked whether I was religious, or Christian. Sometimes that meant saying yes, sometimes that meant saying no, and sometimes that meant saying I would like to be. I always took John Jackson's approach of employing "ethnographic sincerity," which Jackson defined as both a way of being in the world and a way of doing ethnography that treats "other subjects / informants more robustly as fully embodied and affective interlocutors" (2010, p. S285). Affect is part of the experience of fieldwork, and we should embrace it, Jackson has argued, not just as observers but as humans interacting with other humans. Ethnographic sincerity recognizes and honors the fact that "informants embody an equally affective subjecthood during the ethnographic encounter" (Jackson, 2010, p. S281). As such, it becomes a crucial way of practicing ethical research in an age when informants and ethnographers are increasingly in contact with one another in various aspects of professional, personal, and political life. This book is not about me. It is about the believers and media makers whose stories I tell. But, although it was not my intent, and it came as a total surprise, I can sincerely say that meeting, reading, and listening to my informants changed me.

CHAPTER OUTLINES

Chapter 1 explores the digital tools that have come to define evangelical church spaces in the past ten years. I situate these changes within the history of the so-called "church growth" movement that has seen charismatic preachers garner tens of thousands of followers that have met in stadiums, in multisite churches, and online. I explore how the drive to continually integrate more technology into evangelical church services is the result of a strategy that has focused on catering to the aesthetic preferences of white suburbanites. I argue via Life.Church, the "start-up church," that this trend has continued as evangelicals adopt digital habitus.

Chapter 2 is centered on Silicon Valley. Through interviews and ethnographic work I explore the attitudes of Christian founders and tech workers. Although Christian businesspeople believe in the redemptive potential of technology and business, they struggle to define their place in tech culture. And, I argue that in their negotiations with tech culture, these entrepreneurs also fail to reckon with the biases of the technology industry and instead often reproduce them in their own businesses.

Chapter 3 looks at how a network of digitally minded Christians are trying to innovate and hack digital tools to make them suitable for missions work. These Christians are grappling with a set of assumptions about the portability and cultural neutrality of digital tools, and those in the "mobile ministry" movement, inspired by free culture advocates, have tried to imagine ways to spark indigenous media production, much as the tech industry often purports to do—although their efforts have often laid bare the biases endemic to Western-based technologies and the ideologies they carry with them.

Chapter 4 explores Christian influencer culture. I analyze how prominent Christian women have used social media to garner charismatic authority, and I look at the new issues and publics that they have mobilized. I argue that the new forms of feminism taking shape in evangelical culture are restructuring evangelical authority with their own, potent forms of online activism.

Relatedly, the uprisings in 2020 forced evangelicals to face their racial biases, and in some cases, their racist history. Chapter 5 looks at how prominent evangelicals reacted to the uprisings and explores how the Black Christian podcast scene became a space for robust discussion about race and racism in the evangelical church. Black Christians who have been subjected to racial trauma in white evangelical spaces are voicing their dissent and urging the white evangelical church to meaningfully atone for its racist past. Into the crises of summer 2020, when a pandemic raged and American cities burned, Black Christians took to podcasts and used the affordances of this medium to outline a path forward for the Christian church.

This book explores the American evangelical preoccupation with consumer technologies and focuses on how one prominent American subculture is changing as they embrace them, but it is not only about evangelicals, it is about our shared digital culture, and our faith in, reliance on, and suspicion of the rapidly advancing technologies that increasingly surround us. In the end, if American evangelicals are changing as they respond to a ubiquitous digital culture, it may be because we all are.

1   The Church

FROM THE MEGACHURCH TO THE START-UP CHURCH

I'm sitting in the airy lobby of a Baptist church outside of Nashville, Tennessee, where I have been told that it is a minor scandal when the pastor does not to wear a tie on a Sunday. I'm enjoying a cookie from the in-house coffee shop and talking to the church's digital strategy director, whom I will call John. John tells me that he fears for the long-term viability of evangelicalism. Church leaders just are not keeping up with advances in American culture, he tells me. The church is getting grayer, and the people in charge are not paying enough attention to the interests of younger generations. "I'm going to tell you that God's Word is relevant for your life," he begins, explaining how young people experience church. "I'm going to tell you what He has pertains to what's going on right now—but everything I surround you with in that room you walk into is not relevant to your life. At home you experience five screens while you're watching TV. We're not doing that at church." For John and others the problem with church and with evangelicalism as a whole is that it is not attracting a younger generation attuned to media technologies and the entertainment value and social engagement they offer.

John has worked in Christian institutions most of his adult life, and although he voices his concern more forcefully than others, most of the people I spoke with for this book expressed some version of this same sentiment: the church is not keeping up with popular culture, and if it continues to be oblivious to the technological changes happening in the world around it, the digital habitus that has come to define the American middle-class experience, it will disappear. There is some statistical evidence that this is true. The growth of the millennial "nones"—those who claim no religious affiliation—is often cited by Christians as a particularly troubling problem.[1] For many evangelicals the way to attract these younger people is by paying attention to changes in the technological landscape. Their message, as they see it, is timeless, and if the right tools are employed to express that message, the church will thrive. If evangelicals can harness the power of the smartphone, the VR headset, the Apple Watch, perhaps they can reenergize evangelical culture and generate revival. But, if they can't, some evangelicals fear evangelicalism might fade into history. For many, this means churches must embrace technology within their walls, and that they should learn from the workings of those tech companies that clearly have a hold on the popular imagination.

As my conversation with John continues, he explains that the problem is not just that church is not entertaining people. It is that church no longer understands how people think and operate in the contemporary world. He believes churches should think about their parishioners in the way that companies like Google and Facebook think about their customers. "People have a deep desire to be known," John tells me, "and if people are giving over these massive amounts of data on Facebook and on other platforms, churches should coalesce that data.... We have to think in such a way that we really individualize stuff for people." For John, the path forward should be moving the church into the present by understanding the digital habitus of young people conditioned by new media companies who use individualized targeting strategies to serve them content. He wants churches to integrate these practices into church services and church outreach. His concern, however, is that the church as a whole is not up to the task.

Adapting church services to changes in popular culture has been a central preoccupation in evangelical church leadership since the "church growth" movement of the 1970s. This strategy relied on catering to the habitus of the American middle-class suburbanite, and it proved successful in establishing a network of megachurches that came to define evangelicalism for a generation. Although there are myriad versions of evangelical church, there is also a normative model for how successful churches should look in evangelical culture. For years the megachurch has provided that model. This chapter charts how churches are beginning to cater to digital habitus and in so doing are generating new aesthetics and new liturgical modes. I argue that Life.Church[2] represents the apotheosis of a large, dispersed network of churches that have attempted to integrate technology and digital habitus into the spaces of their churches and the strategies that define their outreach. For many evangelicals this has become the normative model of how successful churches should look.[3]

Life.Church calls itself a "start-up church," and that moniker has been reinforced by their very successful app, the YouVersion Bible App. In 2012 the *New York Times* reported that YouVersion was the result of a multimillion-dollar investment that the church made in technology. In 2013 alone, Life.Church reportedly spent $20 million on the app (see O'Leary, 2013). Beyond the Bible App, however, Life.Church has provided resources and guidance for churches that want to move into the digital age, and their success and growth indicates that many evangelicals do. This chapter discusses how and why enthusiasm for new media technologies has taken hold in evangelical churches. I argue that this is a continuation of the "church growth movement" focus on the habitus of the American middle-class consumer. But even as the proponents of technological modernization in the church argue that their work is moving evangelicalism into the future, they are still refusing to face the inequities around race and class that their focus on digital habitus both reflects and generates.

CHURCH CREATIVES, CHURCH GEEKS

Vintage video games are set up in a large, loftlike room with exposed beams and industrial-style fans. Bearded men wearing stylish T-shirts and jeans cradle iPads as they wait for their turn at the nearest console. This building and its inhabitants look like they could be in Williamsburg, Brooklyn, the epicenter of hipsterdom, but they are standing in the lobby of Watermark Community Church, an evangelical megachurch in Dallas,

Texas. This is the scene at Echo,[4] a yearly conference that promises to be the meeting place for "artists, geeks and storytellers who serve the church" (EchoHub, 2018).

The theme of the 2013 conference I attended was "8-bit"—an early form of computer animation, so named because it only allowed for 8 bits per pixel. This animation was developed in the late 1980s and used in video games like Space Invaders and Super Mario World. The conference organizers used 8-bit in all of their promotional materials and created Nintendo-style 8-bit animations that were accompanied by bass-heavy electronic music to introduce each of their main speakers. The Echo organizers chose this theme because, as one explained in his introductory remarks, his generation of Christians came of age spiritually at the same time as video games were coming of age technologically. Thus a generation of millennial evangelicals share a collective memory rife with coexisting Christian and technological narratives, which was externalized in Echo's promotional materials and conference stylings. The 8-bit imagery that pervaded the event signified the personal engagement with media that members of this group overwhelmingly had in common. At this conference those who identified as "church creatives" or "church geeks" (both of these terms were used at the conference but are also used in broader circles) hailed each other and reinforced their social bonds. Many people who had followed each other on Twitter for years shook hands for the first time at Echo. As the conference organizers posted on their website: "Echo has created a tribe . . . and one that we love. We love connecting with hundreds of like-minded creative types in the Church. Echo has become an annual reunion of sorts for us and many others. We deeply cherish this reunion and the friends we get to see" (EchoHub, 2018).

So what is this tribe whose solidarity was on display at the Echo conference?[5] Who are these church creatives and church geeks? The people who came together at Echo are the vanguards of a new generation of evangelical church professionals that hope to push the church into the digital era. As churches have evolved into the high-tech, multimedia spaces that have come to define evangelical worship, these new careers and cultural figures whose authority rests on their ability to understand digital habitus have emerged as central to church leadership. But even though they are young and hip, the ideals that drive them are the same

ones that have guided evangelical churches since the 1970s and the founding of the church growth movement.

MAKING THE MEGACHURCH

Although the earliest evangelical churches in the United States emulated
the style of European cathedrals, as American evangelicalism's democratic,
populist leanings began to dominate, evangelicals experimented with different forms of church. Church services in theaters became popular (see
Kilde, 2002), and the Second Great Awakening saw preachers using out-
door spaces to energize mass gatherings. And as evangelicals adapted to
an audience of American Christians who seemed to crave a flexible form
of liturgy, they were rewarded with congregants. In this way, through continual iteration, evangelicalism in America evolved as a populist religion
with a strong bias toward shaping its culture and liturgy based on what
proved popular in secular culture.

This willingness to adapt formal structures to the preferences of parishioners and potential parishioners reached an apex with the church growth
movement, which began in the late 1970s and saw megachurches pop up
in suburbs across the country. Megachurches have been defined as those
churches that house two thousand or more parishioners per weekend
(typically churches offer several services between Saturday and Sunday).
In their "Megachurch Report" (2020) Warren Bird and Scott Thumma
surmised that the median megachurch had twelve hundred seats in its
auditorium. Megachurches also host many "small groups" in which more
manageable groups of parishioners come together to do Bible study or
otherwise create small community spaces—such as men's groups or cou-
ples' groups—that are meant to counterbalance the largeness of church
worship.

As megachurches began to spring up in the 1970s, "church growth con-
sults" proliferated and a suite of books were (and continue to be) written
on the subject. The Christian publisher Zondervan has a "church growth
section" on its website that contains more than a hundred titles. At its
core the church growth strategy focuses its outreach on "seekers"; seeker-
churches want to grow their church by converting previously unaffiliated

or otherwise affiliated people. As sociologists of religion have asserted, after the 1950s patterns of American religious affiliation changed (see Wuthnow, 1998). It was no longer assumed that one would follow the religious traditions of their family or community. Instead, the popular focus shifted to individual notions of spiritual progress and reward. Wade Clark Roof (1999) has called this America's "spiritual marketplace,"[6] a field in which religious producers use various strategies to attract spiritual "seekers."[7]

Rick Warren's 1995 classic evangelical church growth manual *The Purpose Driven Church*, which has sold more than a million copies, maps out how churches can attract these American seekers. Warren's strategy relies on catering to specific demographic and cultural norms that he gleans, as a business would, using survey research and census data. Warren's initial goal was to plant a church that appealed to middle-class suburbanites in southern California. He wrote about and diagrammed a composite figure that would fit into his imagined church. This was "Saddleback Sam." In his book Warren explains that Saddleback Sam (named for Saddleback, the community in Orange County, California, where Warren wanted to plant his church) was a hard-working, well-educated, professional person who had some apprehension about organized religion. To reach a person like this, Warren thought, churches had to cater to their needs and develop spaces where they would feel comfortable.

As Warren saw it, Saddleback Sam liked the strip malls and big-box stores peppered around his neighborhood, so churches should think about the aesthetics of those stores when they crafted their spaces. "When my friend Larry DeWitt was called to pastor a church in southern California," Warren explained, "he found a small clapboard church building in a high-tech suburban area. Larry recognized that the age and style of the building were a barrier to reaching that community. He told the church leaders he'd accept the pastorate if they'd move out of the building and start holding services in a Hungry Tiger restaurant. The members agreed" (Warren, 1995, p. 269). For Warren, the traditional style of the central steepled church did not appeal to those suburbanites who saw it as anachronistic. Instead, they cast the familiar environs of a corporate chain restaurant as a more appropriate setting for a church. Warren's story illustrates one of the tenets of church growth: to grow, a church must tap into the style and culture of a community. "To penetrate any culture," Warren wrote, "you must be willing to make small concessions in matters of style in order to gain a hearing" (1995, p. 196). For Saddleback Sam, old church buildings carried the baggage of what many viewed as an oppressive religious past, while malls, theaters, and restaurants spoke to the consumer-driven present of the American suburbs. In this way, Warren was urging pastors to use the aesthetics and class habitus of the suburbs to attract parishioners. But notably, Saddleback Sam, as pictured in Warren's book, is white. The imaginary of suburban life, what Donald Trump once called "the suburban lifestyle dream," has presented a vision of respite from the city that expressed a (typically) unspoken racial vision.[8] As white populations abandoned cities and moved into the suburbs in the 1950s, evangelicals isolated themselves from other races in metaphorically closed communities. So the church growth movement catered to a suburban habitus that was assumed to be white.

Writing about the practice of photography in the 1960s, Pierre Bourdieu described how "the most trivial photograph expresses, apart from the explicit intentions of the photographer, the system of schemes of perception, thought and appreciation common to a whole group" (1965/1990, p. 6). The aesthetics of photography are established by social groups and understood within them. So too with church styles that became common and popular in the suburbs. In attempting to remain "relevant"—a term evangelicals often use to describe an ideal vision of an engaged church culture that can communicate with a modern population of churchgoers—evangelicals created spaces that attracted a certain type of person. Megachurches thus became classed and raced spaces, communicating cultural norms through their aesthetics. This is one of the central reasons that American evangelical churches have remained largely segregated spaces (see chapter 5 for a deeper dive into this subject).

Evangelicals have tried to port their central church growth strategies to other contexts, for example to big cities, but they tend to face a host of problems. As one Black Christian has written, "There are plenty of great books about urban or diverse church planting, but they are mostly written from a cultural and privileged bias. They write about the complexities of planting churches in cities, but they ignore the complexities of contextualization in specific ethnic communities. They often exclude certain minority groups. So even our progress made in church planting in urban cities results in white churches being produced in ethnic and economically diverse cities and cultures" (Holmes, 2016). In general, the popular aesthetics of the church growth movement attract the Saddleback Sams of the world, and they create comfortable spaces for him. Christian author Latasha Morrison (2019) has written about the alienating experience of being a Black woman in a primarily white evangelical church, explaining that "some of my white friends thought color shouldn't matter in the body of Christ, an easy thing for them to say. I'd ask them to imagine themselves in an all African American context, attending services where they never heard music by Hillsong, Bethel, Chris Tomlin, or Elevation Worship, just to name a few. Wouldn't that create a cultural shock?" (p. 15). Here Morrison references the culture of evangelical churches where music from Hillsong and others appeals to a white audience but often leaves nonwhite congregants feeling alienated. White congregants, she explains, do not accept that their cultural products might not be universally appealing.

It is rare to hear white evangelicals interrogate the reasons why their churches have remained racially homogenous and that may be because these strategies have worked to attract large populations of followers; the fact that those congregations are not diverse is not considered a problem.[9] In other words, if it ain't broke, don't fix it. In Warren's diagram Saddleback Sam is carrying a mobile phone and has a pager attached to his belt. In 1995, when Warren wrote *The Purpose Driven Church*, these technological accessories indicated that Sam embraced an "early adopter" mind-set. Because of this, the church growth movement highlighted the technological tools a white suburbanite would feel comfortable with and the tools that would set their church apart from the old-style, stodgy, or mainline churches of Sam's past. And they integrated these into their services using expensive screens in the church worship space to signal to the Saddleback Sams and Samanthas in attendance that their church was modern, contemporary, and relevant.

In her study of megachurches, Jeanne Halgren Kilde (2006) explained that "megachurch auditoriums not only include stage areas designed to accommodate large screens—Grace Church, for instance, has two huge screens on either side of the stage—but also eliminate all natural light from the room to optimize video clarity" (p. 243). Screens became the focal point of the worship space in evangelical megachurches, and as such they have more prominence than crosses or traditional religious imagery. In sometimes stadium-sized megachurches, like Joel Osteen's Lakewood Church, which houses more than forty thousand congregants over the course of a weekend, screens are a necessity. But the fact that in the megachurch "the role of congregants in worship ritual consists primarily of watching screens" (Kilde, 2006, p. 244) is also an extension of the seeker-focused strategy that realizes, as Rick Warren (1995) asserted, "television has permanently shortened the attention span of Americans" (p. 255). Because television has become the primary mode by which suburbanites interact with culture, evangelical churches understand the power that screens hold for them and thus they employ screens as a means to attract these spiritual consumers. The experience of church, both in personal worship (singing along with the worship band on stage, repeating prayers after the pastor) and community worship (the understanding of a collective experience), is mediated through large screens at the front of the auditorium. And this fact also requires that church geeks be available to service this technology.

Digital technologies allow for continuously changing content, and producing this content becomes one of the main concerns of megachurches. Northland: A Church Distributed, the former church of megachurch pastor Joel Hunter, produced a documentary that explains how they prepare for weekly services. This documentary reveals that Northland focuses just as much on the technological setup than on the substance of the message. As one church leader explains: "We let a voice-over and scripture and the music and the lights and the video kind of tell the story" (Northland, 2011). "Telling the story" in a contemporary megachurch does not only mean crafting a sermon, it means producing an entertaining and technologically advanced service that can keep the attention of parishioners who have been trained by mass-media environments. In the documentary the crew follows Northland's "lighting/staging director" and shows him renting a fog machine. Throughout the film church leaders have discussions that focus on "creating an environment" for the audience.

In the Middle Ages the Catholic Church created an environment for their churchgoers, making worship into a ritualistic event that was meant to transport the individual into the realm of the sacred. Conversely, in the megachurch church leaders borrow on the familiar paradigms of the secular entertainment industry ostensibly to achieve the same end. Throughout the service the megachurches' screens display the words to worship songs, the text of Bible verses, and play video to introduce segments. Megachurches understand that they are preaching to an audience that expects entertainment, expects production values, and expects screens— this is where "church creatives" come in. To be successful, megachurches need people who can produce what parishioners will see as sophisticated media content. Churches hire MFAs, they employ people who have worked for production houses and television studios. Not unlike the days of rose windows, evangelical megachurches have to hire artists—these "church creatives"—who have a familiarity with the entertainment industry and a knack for storytelling. But like all good storytellers, these artists also understand their audience, the same suburban audience Saddleback Sam was meant to represent. Because of this, evangelical liturgy—from the media to the message—continues to attract and target those middle-class American seekers that have proved receptive to these strategies.

THE MULTISITE CHURCH

The seeker-church strategy relies on continual change and flexibility as a means to attract new initiates, but as megachurches grow, logistical problems can limit growth. Books on church growth note that parking becomes an oft-lamented issue and seating becomes another. One way that evangelical churches have escaped this problem is through establishing satellite churches in other locations, thus enabling megachurches to become "multisite" churches. The majority of megachurches are now also multisite churches.[10]

"Multisite" means different things in different settings. Some multisite churches hold concurrent worship services in which video from one site (proponents advise against using terminology such as "main site") is simulcast into another or many other sites. Other churches transfer their brand of worship to various locations through nonsimultaneous video teaching. And still other multisite churches develop teaching programs to train new pastors in their style of preaching and revamp old sites with their style to attract new followers in new locations. One proponent of this model explains that "the multi-site movement is a strategic response to the question of how to maintain momentum and growth while not being limited to the monolithic structure of a megachurch" (Surratt et al., 2006, p. 7). Multisite churches extend the reach of a successful church's style or brand. And because many churches rely on the charisma of a lead pastor—for example, Rick Warren of Saddleback and Craig Groeschel of Life. Church embody this charismatic preacher archetype and run multisite churches—multisite churches allow a megachurch to extend the reach of a charismatic leader's particular brand of preaching. Some multisite churches have locations in many states, others even boast satellite campuses in other countries. On pastor notes *"multi-site* summarizes today's approach to church in which geography is no longer the defining factor" (Surratt et al., 2006, p. 27, emphasis in the original). By going multisite and overcoming the restrictions of a single building, churches rely on media, technology, and, again, a cadre of church geeks to facilitate these increasingly important media experiences.

Joel C. Hunter, the aforementioned leader of the Florida church known as Northland: A Church Distributed, wrote a manual outlining how and why his church adopted the multisite model. In it, he expressed a commonly voiced frustration with the narrowness of the local church, casting the building of the church as a binding space, one that confines bodies and, by extension, the minds of the parishioners. He explains that "we will miss so much if we limit our exuberance to what happens within the walls we've built" (Hunter, 2007, p. 47), writing that the traditional model runs the risk of being "enclosed and self-limiting" (p. 23). To solve these problems, Hunter established multiple satellite locations all connected together to enable concurrent worship services. He justified this move by framing his understanding of a "networked model" of church as similar to the trinitarian nature of the Christian God who is in Christian theology simultaneously three beings: the father, the son, and the Holy Spirit. In his manual Hunter (2007) explained the earliest iteration of this model:

> The worship services began with T1 phone line hook-ups for video and audio capabilities. Then, because of the geographical proximity we laid fiber-optic cable between the two worship sites, enabling us to worship interactively in real time. We have responsive readings duets, and other types of worship leadership exchanges between the two sites. There is truly a feeling of

togetherness. On occasion, I start my sermon at one location, only to finish
it at another. (p. 41)

Interactive technology allowed Hunter's megachurch flexibility. No longer were the spatial realities of the church building a concern. As megachurches become multisite and innovative technologies open up new
frontiers, evangelicals see themselves as defeating the limiting category of
space. They theologize their practices, as Hunter does, connecting them to
biblical stories, metaphors, and analogies.

At the root of the enthusiasm underlying the multisite model is the idea
that church is a scalable project. Consider one evangelical response to the
multisite movement: "The possibilities are limitless, especially with contemporary technology" (McManus, 2006, p. 8). Proponents of this model
believe that church can expand in geographical reach without losing its
essential purpose as a place of congregation, community, and worship.
Anna Tsing (2012) has discussed scalability as a central ideology of contemporary capitalism, and as in the megachurch, evangelicals have looked
to corporate ideals for guidance on how to scale churches in the manner
of chain restaurants. One how-to manual on multisite churches mentions
the Holiday Inn and Krispy Kreme Donuts as models churches should
emulate (see Surratt et al., 2006). And at two of the conferences I attended
there were several presentations devoted to creating and maintaining a
church "brand" that drew on influences from the corporate world such as
Starbucks, Costco, and In-N-Out.

As in the megachurch, the assumption is that corporate models have
proven successful with the middle-class consumer, and thus churches
should use them to attract this type of person. And as with the aesthetics of the megachurch, the reliance on corporate models is rarely interro-
gated. Corporate business models that successfully excite consumers must
be doing something right, the logic goes, but this way of thinking does not
account for the values that these business models port along with them.
Relatedly, the multisite model relies even more on media production than
the megachurch. The most successful multisite brands are those that are
also media producers, like the über-popular Australia church Hillsong,
which has churches in twenty-three countries and claims to have 150,000
members. It is their brand of original Christian music, which they produce and distribute throughout the world, that has sparked their global

popularity and made them a favorite of Christian celebrities such as Justin Bieber.

But even in more humble churches, multisite requires that they produce video sermons that can be broadcast in other sites, so evangelical pastors become modern-day mini-televangelists. One multisite pastor expressed the anxiety this can breed: "It is also a challenge to feel like a pastor in a church that you never see and that only sees you on video" (MacDonald as quoted in McConnell, 2009, p. 22). But this model remains popular in evangelical culture despite the drawbacks that come with growth, because it has allowed churches and charismatic pastors to extend their reach, sometimes even globally.

THE ONLINE CHURCH

Doug Estes's book *Sim Church: Being the church in the virtual world* (2009) begins with a description of traditional church models:

> Each one has a building with a front door that you open; each one has peo-
> ple who shake your hand; each one has pastors, ministers, elders, or leaders
> who proclaim God's Word to you; each one is real, tangible, physically pres-
> ent. There are differences, but there are more similarities. (p. 17)

Both the megachurch and the multisite church fit this description. But now, Estes (2009) explains:

> A change is occurring in the Christian church the likes of which has not
> happened for centuries. At the beginning of the twenty-first century, the
> church is beginning to be different not in style, venue, feel, or volume but
> in the world in which it exists. A new gathering of believers is emerging, a
> church not in the real world of bricks and mortar but in the virtual world of
> IP addresses and shared experiences. (p. 18)

Sim Church charts Estes's experience as a pastor of a church in the then-popular online game *Second Life*. Estes believes that Christians have an unprecedented tool at their disposal that can allow them to rethink the way church is done. This is much like the model of church in the Book of Acts, before Christian worship was formalized, Estes explains, voicing a common theme that came up in my interviews. The sense is that the internet both allows for something that has never been tried before, and that it represents a more biblical and ancient model of Christianity than contemporary forms of worship. Many people based this exegesis on an understanding of the Apostle Paul's approach to creating and shepherding young Christian church communities through letters or epistles. In their view, Paul used the media of his time to grow the church, so too should evangelicals use media as a means to attract new Christians and create Christian communities.

Online churches have existed in various forms since the 1980s and 1990s. Experiments with online worship have come from many denominations and countries, including the United States, Germany, and South Korea, but it was Life.Church (then LifeChurch.tv) that first debuted the "Internet campus" model of online church in 2006, which has become the predominant form of evangelical online worship.[11] As Tim Hutchings has written of Life.Church's online church, "Church online relies on centralized production of high-quality video resources, including the message of a popular preacher and new music from skilled performers" (2017, p. 200). In his ethnography of various forms of online churches, Hutchings notes that the Life.Church model has proven especially effective at attracting and keeping parishioners (2017, p. 253). And this model has been used by churches all over the country. Indeed, in 2020, 54 percent of megachurches reported that they host an online campus (Bird &Thumma, 2020, p. 27), and this number has only grown, especially as the realities of the COVID-19 pandemic forced once resistant churches to take the plunge into the online world.

Proponents of online church see it as a natural extension of the digital habitus in American culture. If suburbanites—the target of megachurches—are increasingly living their lives online, then churches should be online too. Just as church leaders in the 1980s saw that television had captured the attention of their target demographic, they see the smartphone doing the same today. Then their answer was to integrate screens and high-production value media products into their church services, today the answer has been to make church mobile and online. I first discovered online church in 2009 and when I began talking to online church pastors around this time, most of whom were so-called digital natives in their early twenties, I was confronted with boundless enthusiasm for the potential of this new way of doing church. In the early days of internet campuses many people believed that the online church could be a way to create and shepherd engaged communities of Christians all around the world. In this enthusiasm they shared the excitement evident in the above quotation from Doug Estes.

But by 2013, much of the early enthusiasm had been tempered by the realities of church online. At the Echo conference in 2013 an online church pastor explained that her church had bought iPads for all of their bedridden or older parishioners and they had attempted to provide training on how to attend church online. Still, however, they had seen very little result. The older folks were simply uninterested in this type of church experience. In another session I heard a different pastor bemoan the fact that millennial evangelicals, the presumed demographic of church online, were not taking to it in the numbers her church had hoped they would. In 2017, I caught up with an online church pastor who had been one of the most enthusiastic proponents of church online in the early days. By 2017, however, his initial zeal had dissipated. Although he still believed that church online was an important aspect of contemporary church culture, this pastor no longer thought it was the primary tool through which Christian churches might evangelize the globe.

While Estes and many other Christians who were optimistic about the possibilities of digital technology saw church online as a potential revolution in Christian culture, for many churches it has become a useful, and increasingly expected, add-on. Parishioners who usually attend church in person might navigate to the online campus when they are sick, for example. And since the COVID-19 pandemic, church online has become even more integral to maintaining a church's community structure and brand. The version of church online that has been integrated in evangelical megachurches is different than the model Doug Estes hoped to see—the model of the virtual world church. Instead, it mimics other social media and streaming platforms endemic to the web. Online church campuses have become a way for churches to add another site to a multisite megachurch, and again, they rely on screens. This screen—an individual's laptop screen, for example, or an iPhone—can be the entry into a church community. But this screen is more individualized, more personalized than the screens that dwarf parishioners in the megachurch and in multisite churches.

Megachurches with multisite and online campuses have become the norm. This church style is the result of the long history of the church growth movement that has focused on understanding and adapting to the habitus of middle-class suburbanites. But in the twenty-first century that habitus has changed: evangelicals see digital habitus taking over, dominating the attention of would-be spiritual seekers. Life.Church, the so-called start-up church, is an example of how evangelicals are adapting to the digital habitus that increasingly defines their target demographic. And as in the past, the enthusiasm for this model has allowed its proponents to ignore the racial and class biases that come along with technological adaptations to church culture.

THE START-UP CHURCH

In a video produced by the church titled *LifeChurch.tv's Vision and Values*, they state, in red letters: "We are not a megachurch, we are micro. We are a startup church with a mega vision" (Life.Church, 2012). Life.Church considers itself a start-up church, which indexes the history of "church planting" movements and micro churches that are often praised in evangelicalism, but of course, the term also connotes the colloquial understanding of the tech start-up. In this way Life.Church represents another iteration of the evangelical church's zeal for shifting strategies in order to appeal to Americans on the spiritual marketplace. And although many people at Life.Church's central offices told me that they were just stewarding a movement whose spiritual power came wholly from God, Life.Church also employs strategies borrowed from the corporate world and specifically from tech companies to help them succeed. In turn, the fact that Life.Church employs these methods appeals to the middle-class professionals who populate the pews at Life.Church's many locations and who fund the church's surprisingly sophisticated and complex operation. In December 2017, I had the chance to tour Life.Church's central office in Edmond, Oklahoma, and I was struck by how complex and diversified it is. It looks and runs like a Silicon Valley tech start-up, which is not lost on the people who work there who are proud of their start-up church.

Fittingly, Life.Church began in a two-car garage in Edmond, Oklahoma.

As with tech start-ups, churches often emphasize their humble beginnings and these origin stories are often taken as proof-positive that God is
on the side of the church. Their congregation outgrew site after site before
finally amassing the capital to build their flagship church in Edmond. In
the mythology of Life.Church it became multisite by accident. Pastor Craig
Groeschel's wife Amy went into labor on a Saturday and their child was
born on Saturday night. Groeschel had to make a choice: spend the next
day with his wife and new baby or head back to the church and preach
Sunday services. He decided to take his chances and play a video of him
preaching on Saturday at the Sunday service.

   As Groeschel tells it, nothing changed. At the altar call, people still
raised their hand indicating that they wanted to come to Christ. And,
as congregants continued to flock to Life.Church, the church decided to
set up satellite sites where worshippers could gather to watch a video of
Groeschel's service in other areas: first in Oklahoma, then in Texas, Florida,
Tennessee, and New York. In December 2017, when I visited, they had
twenty-six satellite locations across the country, and by December 2020
they had thirty-six. In Life.Church's central offices they have a wall with
pictures of each of their locations hung in the order that they appeared.
What is striking about these photographs is how similar each of the locations looks. This is not an accident. Life.Church has an interior designer
on staff, and they employ a marketing team whose entire purpose is to
make sure that branding remains consistent in all of their multisite locations—again, this strategy emulates the chain-restaurant business model
and takes seriously the idea of a scalable church brand.

   Touring Life.Church's central offices on a bustling Monday, I was struck
by the complexity of the operation. A staff member told me that was due to
the fact that Jerry Hurley was on the directional leadership team (abbreviated as "the DLT") of Life.Church. Hurley, a district manager from the
Target corporation, brought his experience managing a chain of corporate stores to Life.Church. The corporate feel of the church is especially
evident at the central offices. Individual offices have glass walls to indicate transparency and openness. At each staff member's desk is a placard
with a printout of their Myers-Briggs Type Indicator. You might run into
an INTJ (someone with an introverted, intuitive, thinking, and judging
personality type) or an ENFP (someone with an extroverted, intuitive,

feeling, and perceiving personality type), and seeing that, I was told, you would know exactly how to approach that person and how to speak to their strengths. "Speaking to strengths" is important at Life.Church. In order to promote a culture of positivity, the staff is never meant to emphasize weakness. When I asked someone about whether they had gotten criticism from another similar ministry, I was told that they have a policy of never speaking badly about any other ministries. Thus the atmosphere at Life.Church is overwhelmingly positive by design, but beyond that, people seem to genuinely want to be working for the church.

At one point I asked a Life.Church employee if there was anyone working at Life.Church who was not a Christian. He seemed flummoxed by the question and, repeating the mission statement of the church, told me that their mission was to lead people to become fully devoted followers of Christ so...no. Unlike a business, Life.Church does not have to abide by religious antidiscrimination laws that bind, for example, faith-based business start-ups, and this influences the work environment. People come from all over the country to work for the church. Several people told me that they believed they were called to Life.Church by God. And some saw their position at Life.Church as the answer to a prayer. I was told that people came from large tech companies like Amazon and Apple and took a pay cut to work at Life.Church because they believed that they could use their "gifts" in the service of a greater mission there. This church attracts the cream of the "church geek" crop. Life.Church's parishioners know this as well, and it enhances the enthusiasm they feel for their church's mission. One parishioner proudly told me that Bobby Gruenewald, the church's "Innovation Leader" and the central figure behind the YouVersion Bible App, had left a company worth $20 million to work for free at Life.Church. Both the fact that Gruenewald had been a successful businessman and the fact that he had left it behind to work for Life.Church were admirable things in this man's opinion. They proved to him that Life.Church was an important, successful, "relevant" place.

At Life.Church's central offices, Gruenewald leads a team of developers called the "digerati" team, which is what ultimately sets Life.Church apart from other megachurches and multisite churches and makes it a start-up church. Gruenewald and the digerati team create "digital missions," which are meant to be tools that evangelicals can use in the digital environment.

One of their digital missions is "Open." "Open" offers sermon series for adults and children; various media that smaller churches might pair with their sermons; financial worksheets tailored for churches to help them with payroll and tithing; free Christian music; and software and apps for church use, including the platform on which Life.Church hosts church online. Life.Church offers all the resources a pastor might need to start their own church from the simply logistical to the high-tech. On the FAQ section of the "Open" site, they state their purpose clearly: "We give away free resources to churches because we believe that they belong to God and His entire Church. So no need to give us any credit, just give all that credit to God!" (LifeChurch.tv, 2014). Life.Church sees its work as instrumental to an imagined global church community united under the authority of God. When I visited the office for "Open" at Life.Church, I spoke with five staff members who told me that they were experimenting with what they call "Open digerati." This will be an open-source software platform that church geeks can tinker with and adapt—in effect it will truly "open" Life. Church's digital resources.

Life.Church also has an office for their church online team, which fields thousands of online prayer requests every day. This part of the digerati team stewards the online church campus, and they try to grow their online audience by buying keywords for whatever people are searching on the internet in order to get them into church online. In this way, Life.Church mines the internet, using the tools offered to businesses to do targeted advertising, to find people who may be struggling. And because they use the AdWords system, they have the ability to track people through Google's analytics. I asked the online church team about trolls, and they said it was a big part of how they had to manage the church online platform. Because people might see an ad for Life.Church's online church campus when they search for, let's say, "porn" (this is a keyword that Life. Church buys), some are not too happy to be redirected to a church service. Yet they were optimistic about church online's success. They said that at a recent service eleven people had raised their hands (an option that users can click on through the online platform) to become followers of Christ. "That's eleven more people who know Christ!" I was told by an excited Life.Church employee. Although they use Google Ad Dollars to track online parishioners, as this reaction indicates, the stakes for the Life.

Church are different than those of a business. They are not using this data to predict buying patterns, as a for-profit company would, but to track their online parishioners' spiritual progress.

Near the church online team is another office that houses people tasked with coming up with whatever is next in the technological realm. These people (I only saw two men working in that office on the day I was there) sit in a room all day anticipating what might happen with technology and how Life.Church might use new innovations to reach more people. They are trying to come up with the next YouVersion, in whatever form that might take. Like many tech companies, Life.Church institutionalizes future-casting as a means to stay on the cutting edge of technological innovation.

The digerati teams's most successful "digital mission" is the previously mentioned YouVersion Bible App—a free smartphone application with which users can access a digital Bible and connect with friends or churches to participate in Bible study. As of February 2021, the app has been downloaded more than 465 million times and has been translated into every major language.[12] Life.Church created a version of the Bible App for the 1,140 people who exclusively speak Samoan, and they have done the same for similarly underused languages like Huilliche (Chile), Longto (Cameroon), Hupde (Brazil), and Ama (Papua New Guinea). By doing this, Life.Church hopes that their app, and of course the Good News that it contains, can reach every person on Earth. YouVersion has a large space in the central offices that houses programmers, engineers, and designers, set up like the offices of a tech company. Coders sit at desks outfitted with multiple monitors. Engineers' desks are outfitted with a light that is either green or red, indicating whether or not they can be interrupted. I spoke with one man wearing a sweatshirt embossed with the Apple logo, who excitedly told me that he had gotten to go to an Apple event to test the YouVersion app.

That Life.Church uses business strategies borrowed from the tech world in service of their church makes them appear to parishioners and to their own employees as particularly "relevant." And as a leader in the digital church movement, Life.Church has been able to attract church creatives and church geeks from around the country who see their work at Life. Church as purposeful, important, and in many cases directly guided by

God. Through their focus on technology, Life.Church makes the case that evangelicalism is a modern religion with a role to play in the digital world, and it does so by extending the church growth strategies established in the 1970s and 1980s that placed their focus on attracting American middle-class spiritual consumers. And as in the church growth movement, church leaders carve their path in this new terrain by using strategies that emulate those of successful businesses—in the case of Life.Church, tech businesses. Evangelicals assume that something that proves popular in the American marketplace will easily translate to the spiritual marketplace. Because they measure their success based on growth and numbers rather than other metrics, their strategies have been seen as successful, enviable. Indeed, nearly everyone I spoke with for this book mentioned Life.Church as the model that other evangelical churches should follow into the digital age.

But what is it like to attend a start-up church? Next I sketch my experience of attending Life.Church and watching Craig Groeschel preach in person, attending a multisite Life.Church location, and attending church online. Evangelical church spaces have evolved throughout the past forty years—from the megachurch movement, to the multisite, the online church—and Life.Church's strategy encompasses all of these iterations. But Life.Church has also updated the playbook for evangelical liturgy in the twenty-first century to reflect the digital habitus of their target audience.

ONE CHURCH, MANY LOCATIONS

Flagship

"Is this it?" Asks the Lyft driver. Even though I am expecting a relatively nondescript building, I am not sure I am in the right place until I see the two large satellite dishes out front. "This is it" I tell her. This is Life. Church's flagship location in Edmond, Oklahoma, where Craig Groeschel preaches in person every weekend. His sermons go out to twenty-six connected campuses, to "networked churches" ("churches" housed in people's homes, in community centers, or in other locations), and to a large online church audience. Life.Church estimates that about seventy thousand people watch some version of Groeschel's weekly sermon each weekend. On the front page of the Life.Church website they claim that they are "One Church, Multiple Locations" and explain: "A church isn't a building—it's the people. We meet in locations around the United States and globally at Life.Church Online. No matter where you join us, you'll find friendly people who are excited to get to know you!" (Life.Church, 2018).

Outside the auditorium in Life.Church's flagship location are high-top tables and black leather couches for people who came early for the church service to sit and gather. Coffee, tea, and cookies are on offer. The church, as all Life.Church locations, has an industrial aesthetic with polished concrete floors and exposed ductwork. Simply designed text posters on the wall quote Life.Church's central tenets like "Our mission is to lead people to become fully devoted followers of Christ." But other than the references to Jesus in these quotations, the church has very little religious imagery. For example, there is the red Life.Church sign instead of a steeple outside. Volunteers for the church wear red branded T-shirts with the Life.Church logo on them over jeans, and they smile and greet people as they walk in. Inside and out, it looks more like a Costco than like a cathedral. On the day that I attended, the church was decorated for Christmas and several small sets had been constructed with Christmas scenes. The church offered a professional photographer at each of these so that attendees could take Christmas photos with their families. The crowd at this location was mostly white, and judging by the cars in the parking lot, and the self-presentation of the people (including their various technological accoutrements such as iPhones and Apple watches) seemed to indicate that they were middle to upper-middle class.

About fifteen minutes before the service was set to begin, I entered the auditorium, which was set up like a theater with visible insulation lining the walls to enhance sound quality. Two large, broadcast-quality cameras loomed above the parishioners, who sat on folding chairs. And the lighting was kept dim outside of the stage. It felt like we were sitting in a studio, like we were about to watch the live broadcast of a television show. Before the service began, "trailers" played on the three movie-sized screens at the front of the room. On this day, there were two videos that were roughly four minutes in length. The first was a behind-the-scenes look at a new song being produced by the church, and the second was an interview with a Christian singer who uses the YouVersion Bible App and who called it "life-changing." Both short videos had high production values; they could have played on any television network.

The service began with the band playing two praise songs, as is typical of a contemporary evangelical liturgy. Parishioners rose to their feet and were encouraged to sing along (the lyrics played across the screens along with the songs) and clap, dance, or raise their arms. Then the campus pastor came on to the stage, urged the congregation to tithe, and introduced another video. This video showed a tribe in Zambia who were able to read the Book of Acts in their native language for the first time through the YouVersion Bible App. He emotionally said that "Bible poverty" might be eliminated in our lifetime because of the work going on right here at Life. Church. He told the audience, "The prophecy is happening," implying that the biblical prophecy that directs Christians to proselytize to "every tribe and tongue" was being fulfilled with the help of the Bible App. Then he introduced the senior pastor, Craig Groeschel.

A highly charismatic preacher, Groeschel uses his droning voice well, making it rapid and loud when he is trying to emphasize a point in the manner of a hip-hop artist. He darts back and forth across the stage, adroitly and seamlessly looking into the audience, then the camera, then back to the audience. He tells personal stories about working out, about his wife and children, about a sleepless night tossing and turning in bed. He tells a story about hearing of a friend's tragic death during Thanksgiving dinner and going outside to weep and pray. Throughout his sermon Groeschel is careful to reference other Life.Church locations at which parishioners might be watching his sermon live. To this end, he says things like, "Somebody in Wichita say it with me," when asking the congregation to repeat a biblical phrase. In introducing the concluding prayer, Groeschel says, "All of my churches, would you pray aloud." In these ways he hails the congregants in the crowd in Edmond as part of a larger, dispersed network of people all worshipping together. Near the end of his sermon music swells behind him. The service ends with another praise song. I cry and laugh along with Groeschel and with parishioners in the audience during his sermon. I leave the auditorium feeling like I have just had a particularly emotional conversation with a friend, who despite everything remains optimistic, upbeat, and secure in his religious faithfulness.

Although some people brought their own physical Bible, most, like me, used the Life.Church YouVersion Bible App during the service. There were no physical Bibles available in the makeshift folding-chair "pews." On the YouVersion App you can search sermon plans, and I found the plan for the day I was visiting, and I followed along and took notes on my iPhone. The app in some ways gamifies the experience of church-going by offering "badges" to users when they perform certain functions on the app. For example, I earned a "YouVersion Badge" when I subscribed to a reading plan on the app. The app also integrates with iOS functionalities. Because I have installed YouVersion on my iPhone, when I send a text message, my phone offers a widget that would allow me to send Bible verses from the app through text. And in this way, Life.Church's app folds into the digital habitus of its users.

Multisite

The following morning, I saw the same sermon preached at another location, Life.Church's second campus, also in Edmond, Oklahoma. This time my Lyft driver knew all about Life.Church. He told me that he was not one to cry easily but that he cried every time he went to service at Life.Church because of the music and the preaching and the Holy Spirit present in the congregation. He asked if I had heard of the YouVersion Bible App, and he told me that millions of people had downloaded it. Life.Church is just so "relevant," he explained. This location, like the first Life.Church location, had the branded sign out front, and although this one had a large cross outside, the inside was similarly devoid of traditional religious imagery. At this service Groeschel's sermon played live on the large video screens at the front of the church auditorium and it was bookended by appearances from the "campus pastor," in this case a woman named Erin Crain. Walking into the auditorium was the same. The turnout was the same. In-person singers and a band played just as they had the day before. The videos played on the three screens were the same. And again, I cried and laughed along with Groeschel and the congregation.

Outside of the auditorium before and after the service, parishioners greeted each other as they had at the first location. I saw a list of "small groups" offered by this location that people could sign up to be involved in. I spoke with a volunteer who told me that it was great to be able to travel and go to a Life.Church in another location. He had recently taken his family to another networked Life.Church location in Kansas City, where they had seen their pastor, Groeschel, preach on the screens there. He told me when you walk into the Kansas City Life.Church location, it looks just like the location in Edmond that he attends. "Like In-N-Out?" I say, referring to the way that chain locations of corporate stores tend to look the same. "Exactly," he responded.

Online

Life.Church's church online opens with a short "Welcome" video that shows quick crosscut images of people logging on, Groeschel energetically preaching, a band playing on a stage, and the sentence "Our goal is to lead people to be fully committed followers of Christ." Then the live feed starts. The service online begins with music, in the same way it does in Life.Church's physical locations, although the experience is different because the shots of the band are seen through multiple cameras in the manner of a music video, rather than from a vantage point of an audience member in the auditorium. I do not stand up from the music like I would if I was in an auditorium, although I bob my head a bit and tap my feet. I notice that I have not thought about how I am dressed, as I would have before attending a church service. I sit at my desk, watching the feed, as I would with any other online video. I resist the urge to toggle to another tab, and check Twitter, an urge that is consistent with my online behavior. There is an empty heart shape at the side of the screen that when clicked on launches a colored heart animation across the screen. This is a similar functionality to Facebook, which allows "likes" to stream across live videos. I see other people's hearts go by when I assume that they want to indicate that they agree with or like something in the sermon on screen. This serves as a reminder that there are other people watching along with me from various locations.

There is a chat screen next to the live feed, where people labeled "hosts" greet users by their usernames as they come into the chat. In many of my experiences of church online I have found that "trolling" is a persistent problem for church online. Sometimes comments by trolls—that is, people who enter online spaces only with the intent to disrupt them through crass or aggressive behavior—are deleted from the chat screen immediately. But often, the hosts will try to engage with the troll, ask the troll what is wrong and whether they would like the hosts to pray from them. Others in the chat are regulars who greet the hosts cordially. The hosts chat back and forth with the regulars, with the trolls, and with each other throughout the service. Occasionally someone says something like "Okay, I'm going to go full screen now, see you later!" People in chat "sing along" by typing the words to the song they are listening to. Users sometimes post emojis to indicate how they are feeling. The chat screen can be toggled to another in a long list of languages at which point it will immediately auto-translate. This constant chatter is markedly different from the experience of attending a church in person.

And there is no sense of placeness in Life.Church's online church, or any online church I have attended. Instead, the sociality of Life.Church's platform resembles that of Twitter, Facebook, or other social media platforms.[13] As in those spaces there is a different type of sociality at play. People say things to others that they would never say in a face-to-face environment. Trolls are the clearest example of this. Trolls in a physical church space would be either reprimanded by the people around them, or in the case of disruptive behavior, would be ejected from the service by members of the church community. Similarly, parishioners do not tend to speak to each other throughout a church service if they do not know each other. Clearly, there are different social rules governing the online church space as opposed to the physical space.

After the music the online pastor, Alan George, comes on screen. He tells a story about an African refugee coming to the United States that illustrates that church online through Life.Church is a "global community." He encourages online church members to invite people to the church. Then he introduces Craig Groeschel. Groeschel's preaching is engaging, but it is more difficult to pay attention to him. Because church online is located on my personal computer—a tool that I am used to controlling and adapting to my own preferences—toggling back and forth, multitasking, and so on, I do not feel the social pressure to be paying close attention that I would feel in a church auditorium surrounded by fellow parishioners. In the chat, during the service one person mentions that he believes that he is a curse. The online church parishioners in chat try to encourage him and convince him of his worth. This is a common experience in church online that is less present in a physical church, where people do not talk to each other during services and where people generally act in according to established social norms.

Many Locations

Life.Church calls itself one church with many locations and attempts to make the services across three different structures feel consistent. Each iteration has a campus pastor, and throughout the service pastors emphasize that Life.Church is one large, global family, rather than a single building or experience. Life.Church is a brand that makes its parishioners feel comfortable because each location has a sameness to it. That the multisite model borrows from the strategies of chain stores is appealing to middle-class parishioners, who see this as a way that the church is remaining "relevant" in American culture. Several parishioners praised the efficiency of this model and were excited by the fact that the church was borrowing strategy from the corporate world. Like the Lyft driver who brought me to Life.Church and told me that they were "so relevant," the fact that Life. Church borrows business models from the corporate world and from the world of tech in their growth strategy appeals to its congregants, who see this fact as evidence that Life.Church is particularly suited to growth in the contemporary spiritual marketplace.

In their attempt to equate all modes of church service such that to watch Groeschel in person, on a large screen, or on a laptop screen all invoke the same feelings of community and worship, they also encourage digital habitus. Technology is used as a way to excite congregants and make them feel as though they are actionably involved in the fulfillment of a biblical prophecy. And the "life-changing" technology that is most central to the experience of Life.Church is their YouVersion Bible App, which they constantly promote. For many, the app is another piece of evidence that Life.Church is a contemporary, practically focused church. Life. Church parishioners tell me that they believe that their church is in the center of a revival. One says, "God could turn the tap off at any time," but as for now, they tell me, they are growing, they are reaching people, and they are doing it by staying on the forefront of technological innovation.

Life.Church's promotional materials emphasize the potential global impact of the Bible App and of their other technological tools. They make the case to their parishioners that their focus on technological innovation is a way that they are able to evangelize the globe. But it is hard to know whether that is really happening. Given their growth in the United States, it seems that the people that this strategy appeals to are the same people that have populated the pews or folding chairs of the megachurch movement since the 1970s—suburban, middle-class churchgoers. Life. Church's model is the result of decades of evolution in evangelicalism that has been sparked by following business models from the corporate world and increasingly by using technologies and practices imported from these worlds, rethought and remixed for Christian purposes. As such, it continues the strategies of the church growth movement that preceded it. And like that movement Life.Church's model relies on an understanding of what is popular to a certain type of person—a Saddleback Sam or a Life. Church Leo—but at the heart of the strategy is a canny understanding of audience that has developed over fifty years of church growth research. Evangelicals have found a formula that works. By relying on high-tech tools and savvy entertainment techniques, evangelicals have been able to dominate the suburbs. Life.Church's strategies continue that trend using digital tools.

SOCIAL DISTANCING

On Easter Sunday 2020, my family, like many in my home state of California, had been "social distancing" for almost a month. In Los Angeles we were under a "safer at home" order, which discouraged any nonessential travel or movement. Though two weeks before, then President Trump had predicted the country would be opened up and the churches would be full by Easter, the threat of the COVID-19 virus loomed large and churches all over the country had been preparing for weeks to put their holiest, busiest day online. Just as work had transferred to the home space, people were doing yoga at home, churchgoers all over the country were worshipping from home.

I tried to tune in to the live service at Oasis church on my Apple TV but was served an error message. My oldest son, then four years old, began to throw a tantrum because he had been anticipating screen time. Instead, I tuned in to the service on my iPhone and saw a couple playing worship music in their home. The woman sat at a keyboard and her partner sat next to her strumming a guitar. Then, Julian Lowe, the lead pastor, stood in his home and preached his Easter Sunday sermon into a video camera. Although this Easter Sunday represented a sudden disruption of church services, the evangelical church had been working toward this moment for more than a decade. This was the moment when a nationwide network of church geeks and church creatives were called upon to share their expertise. Apps and services that had been created the decade before were marshaled to enable online giving and online services.

This chapter charted the changes in the American evangelical church that led to this strange Easter morning when, perhaps for the first time, a majority of Christians all over the country were worshipping and praying together through their digital devices. As in other industries and organizations, the move to online during the COVID-19 crisis was both sudden and the result of a long migration. Since at least 2006, with the founding of Life.Church's online platform, evangelicals have been drawing from the logic of the church growth movement and expanding into the online world. They have hoped to unite the world in Christian community, but in focusing on digital tools and platforms, they have also highlighted the divisions in the world.

CONCLUSION

It's a hot Sunday morning and I'm in the passenger seat of a car with the air-conditioning blasting. We are heading to the beach. I scroll Instagram and see that InstaChurch is going live. I join the service and laugh, thinking I must be the only one of the two hundred or so people in the stream in a car on the way to the beach. Almost as soon as I say this aloud to my husband, I hear "Oh! This one is heading to the beach! That's the way to do it," coming from the livestream. I'm apparently attending church with many other mobile parishioners on their way to celebrate Labor Day.

Alex Dion Winston runs InstaChurch.Live with his wife. He's young and fit, he wears streetwear and has bleached hair. Winston targets his sermons to "YouTubers, skaters, etc." and he also sells "Amen Merch" on his websites. But if that's not your thing, you can also attend church in VR. Former megachurch pastor D. J. Soto started his VR Church in 2016. Using an Oculus Rift or any other VR headset, you can log on to Soto's church in AltSpaceVR. One church consultant, Jason Caston, told me that he see potential for churches to get their content on Alexa, on smart appliances, and eventually into self-driving cars. He imagines people commuting to work and participating in a church service along the way.

As media forms change and capture public attention, evangelicals change right along with them. They adapt their imagination of what church can be to appeal to the habitus of their parishioners and of potential converts. Church may be experienced in an auditorium where thousands of voices sing along to the words of a Christian rock song flashing on an overhead screen. Church may be experienced in a theater where the smiling, familiar face of a distant pastor beams, larger-than-life, out of a movie screen. Or church may be experienced in a living room, with a laptop, an iPhone, or a VR headset providing the interface. At every turn, evangelicals have created new spaces for worship and community engagement in evangelical culture, and these spaces display new social logics and customs. Increasingly church spaces have been infused with digital habitus, and the church has been somewhat ported into the digital realm.

But this focus on new technology and digital habitus is also exclusionary. Yes, these devices and practices have become popular and widespread, but among what demographics? As evangelical churches shift to make digital habitus central to their worship, community building, and outreach, they double down on their efforts to attract and retain the white, middle-class, American spiritual seeker, and they rarely reflect on why this focus on a particular aesthetic only seems to work with this demographic—perhaps because it works so well.

The rest of this book deals with the many ways that evangelicals animated by this imperative negotiate their place in digital culture and in new media industries and how their embrace of technology is changing the contours of this subculture. In the next chapter I look at those tech companies in the subindustry of faith-tech, many of which provide churches with technologies that enhance their digital footprint and which see the digital habitus in evangelical culture as a business opportunity. The world of faith-tech, like the church world, looks to corporate, business ideals as a means to understand how evangelicals might fit into contemporary culture, but unlike the church world the evangelical businesspeople that traffic in the faith-tech industry have to negotiate their place in a cultural milieu that emulates Silicon Valley. In the negotiations between what it means to be a Christian business and what it means to be a tech start-up, these entrepreneurs give voice to the particular reified place that tech business has in the American imaginary and express their hope that their presence in this industry might help redeem it, and by extension might help to spiritualize the world.

2 The Start-up

THE CULTURE OF FAITH-TECH AND THE PROMISE
OF REDEMPTIVE ENTREPRENEURSHIP

The first thing I notice about this tech incubator that takes up nearly a whole floor of a New York skyscraper is the anxious excitement in the air. Young people in stylish sneakers breeze by in mid-conversation wearing branded T-shirts promoting new companies, I have never heard of. The aggregation of quiet conversations and keyboard clicks in this sprawling open office make it feel like a place where deals are being made and things are happening. I'm here to meet with a young entrepreneur I'll call Jack, who sits in the corner managing a small team of three.

When I met Jack in the fall of 2015, he was wearing leather boots, an Apple Watch with a gold band, and a white T-shirt that revealed an elaborate tattoo on his forearm. He looked like the kind of person that works in the tech industry: young, fit, enthusiastic, and tastefully teched up. He told me that he had cut his teeth in tech at Amazon working in user experience design and had worked at start-ups before deciding to found his own company. He described his faith-based start-up as arising out of a conversation with a venture capitalist who encouraged Jack to use his personal experiences to create a new brand. This prompted Jack to think about the centrality of faith in his life, and he told me that he recalled thinking that he didn't just want to make money or create an interesting business model. He wanted to use the connections he made with users on his app to do important work for the Christian community. As he explained it to me he wanted to:

> build something that is meaningful and impactful. The cool thing about it is for me is . . . we're building a community which we can then do great stuff with. We can feed kids every month like we're doing now or as the community grows and we listen to them we can figure out where are those other gaps how can we serve them better with technology and applying it to their faith which to me is the long-term goal. Let's figure out where those gaps exist and try to fill them in the best we can and see how that can impact individuals and hopefully impact other things as well.

Here, Jack adroitly mixes Silicon Valley buzzwords like "impactful" and "meaningful," with evangelical argot like "community," "faith," and "serve." He characterizes his company as another app that might help "save the world," a particular preoccupation of tech start-ups, and he says that his app can do that by strengthening the Christian faith of its users. In many ways Jack's vision for his start-up is illustrative of the world of faith-tech. "Faith-tech" is a space in which Silicon Valley ideologies blend with Christian visions of technology, business, and the future. In this space Christian entrepreneurs eager to bring their faith into the twenty-first century and attract a generation of spiritual seekers defined by digital habitus are creating technological tools for smartphones and VR platforms.

Much like the Christian imperative to be "in but not of the world," Christian tech start-ups are in but not of Silicon Valley—sometimes literally sitting in hubs apart from Silicon Valley as in those communities of tech entrepreneurs in Nashville, Dallas, and Atlanta—which raises a set of questions about faith-tech business: How do Christians imagine the tech sector, and how do those working in it see their work? How does a company that targets a Christian audience and is run by Christians run differently than any other company? What are the big and small concessions these Christians have made to fit into Silicon Valley's cultural milieu, and how have these values influenced and been influenced by evangelical culture?

This chapter is based on four months of ethnographic fieldwork at a faith-based start-up in Los Angeles as well as interviews with founders and CEOs of faith-based companies and venture capitalists who focus on Christian start-ups.[1] Most of the start-up founders I interviewed were running small businesses that had been in operation on an average of two years; however, one founder had been in the faith-tech sector for more than twenty-five years. Most of the start-ups were in the app business, meaning they were working to create smartphone applications; four of the CEOs I spoke with had founded VR companies and were creating content for the burgeoning virtual reality scene. Most of the start-ups I looked at had very few employees, but one had three hundred. Faith-tech businesses sometimes sprung out of the church world—for example, Tithe.ly, a successful tithing app created by a former pastor. Others, like Jack's start-up, came from the tech world. My sample reflects the continuum of businesses in faith-tech, yet what came to the fore in my conversations with these businessmen—and they were with only one exception men—was a strong belief in the power and influence of business and entrepreneurship in American culture similar to that found in places like Life.Church, the start-up church explored in the previous chapter. Yet although Christian entrepreneurs often spoke of the laws of markets as though they were natural laws, they also challenged many of the cultural norms of tech business.

The discourses circulating in the faith-tech sector reveal an evangelical imaginary of entrepreneurship and technology influenced by the techno-utopianism of Silicon Valley. This is clearly expressed in the philosophy of "redemptive entrepreneurship," an idea that prompts Christian entrepreneurs and businesspeople to think through how to Christianize digital habitus, how to support churches entering the digital age, and how to individualize Christian experiences, rituals, and liturgies so that they fit on a smartphone. Their work, many entrepreneurs in faith-tech say, is more than just business: it has the power to be "redemptive" to the tech industry, to digital habitus, and to the world as a whole. But even as they claim to redeem high-technology, faith-tech entrepreneurs have accepted the ideologies of Silicon Valley and have had to adapt to its cultural norms in a way that replicates the problematic blind spots of the American tech industry.

MAKING FAITH-TECH: SILICON VALLEY IDEOLOGY MEETS EVANGELICAL ENTHUSIASM

To understand the faith-tech space, we first need to understand how Silicon Valley's counterculturally inspired aims of social revolution serve as an ideal harbor for Christian entrepreneurs, and how the industry's historical shortcomings are replicated by industries and institutions who emulate it. As Fred Turner (2006) has documented, the roots of Silicon Valley culture go back to the 1960s. During this era Stewart Brand, the radical thinker behind the *Whole Earth Catalog*, inspired by Marshall McLuhan, Buckminster Fuller, and others, created a counterpublic that believed in the world-changing potential of the early internet (Tuner, 2006, p. 89). These "New Communalists" had a romantic vision of the potentials of the new medium and injected a sense of play into computing, which had previously been dominated by an understanding of the managerial and business possibilities of the computer.[2] Their vision of computing ultimately went mainstream when the personal computer entered the consumer marketplace in the 1980s.

In *Silicon Valley Fever* (Rogers & Larsen, 1984), a portrait of Silicon Valley written in the midst of this heady time, the authors painted a picture of the Valley as a place where capitalism in its purest form thrives: "Meritocracy reigns supreme in Silicon Valley," they note (p. 139). In this business world there is a "sense of power of the future" (p. 23). As this description shows, a mystique was beginning to surround the industry in the 1980s. This aura especially shrouded the new poster boys of the personal computer revolution. The American media characterized Steve Jobs and Bill Gates as young antiestablishment figures whose countercultural affectations set them apart from other businessmen of the time.[3] And these tech businessmen continued to assert that their products could improve the world, or in Jobs's iconic (and appropriately ambiguous) phrase, could "put a dent in the universe."

In the 1990s, as the internet transformed into the World Wide Web, companies like Apple, Microsoft, and America Online proved the tech sector was no longer simply the domain of iconoclasts and hackers, rather it was crucial to American economic progress (see Schulte, 2013, pp. 83–112). It was during this time that the countercultural beginnings of the web were married to libertarian ideology perhaps best exemplified by *Wired* magazine, an outlet that extolled neoliberal ethics but remained cloaked in the aesthetic signifiers of youth and hipness (see Streeter, 2005; Turner, 2006, pp. 207–236). Relatedly, as tech businesses grew, they began to adopt what Richard Barbrook and Andy Cameron (1996) have called "the Californian Ideology," which they asserted "promiscuously combines the free-wheeling spirit of the hippies and the entrepreneurial zeal of the yuppies" (p. 1). This ideology is an inherently optimistic attitude that assumes that the world's problems could be ameliorated with technological, market-based solutions—and without the unnecessary intervention of government entities. Those who ascribe to the Californian Ideology see a techno-utopian future that allows for a denial of the structural inequalities that are necessary to produce technological products—for example, the brutal realities of the global supply chain. This is why Barbrook and Cameron (1996) characterized the Californian Ideology as "an optimistic and emancipatory form of technological determinism" (p. 14).

Also in the 1990s, investors flocked to companies with dubious value other than the dot-com at the end of their domain name and the stock market briefly soared in what became known as the dot-com bubble. Thomas Streeter (2011) has characterized this era of speculation as fueled by romanticism and a new imagination of the potentiality of computing and the internet's role in society: "Change the world, overthrow hierarchy, express yourself, *and* get rich; it was precisely the heady mix of all of these hopes that had such a galvanizing effect" (p. 133). Importantly, this bubble was not only fueled by the promise of financial success but also by an understanding of high-technology businesses as particularly unique and special in the American imaginary.

Although the dot-com bubble eventually burst in 2003, the sense of sublimity was transferred to the new technology companies cropping up and rebranding as "Web 2.0." This moniker referred to all of those companies that were able to harness and monetize the content creation and sharing aspect of the web—ventures such as Facebook and Twitter soared. Scholars began to write about the potential of a new participatory culture in which those who were once merely consumers of hierarchically produced media content could actively create or remix content, and some even asserted that the ideals of deliberative democracy could be realized through networked political participation.[4] That Silicon Valley's understanding of itself as a world-changing industry had taken hold in the American imagination was particularly evident during the beginning of the Arab Spring in 2011, which was initially dubbed "the Facebook revolution" in the U.S. media. Many scholars criticized this label as it seemed to simply repackage Silicon Valley technological determinism and sloganeering rather than interrogating the historical and cultural factors at play in the protests that swept the Middle East.[5]

In 2008 Steve Jobs, then CEO of Apple, created the App Store and allowed third parties to create applications for the iPhone, a move that inaugurated yet another tech boom. This development led to the explosion of tech start-ups and mini Silicon Valleys (Silicon Beach in Los Angeles, Silicon Savannah in Nairobi, etc.) all over the world (see Merchant, 2017, pp. 148–184). The app economy fueled investor excitement and led to top-dollar valuations for apps like Instagram and Uber. Apps in the aggregate have "disrupted" many social and economic norms—for example, Uber and Lyft have crippled and contested the traditional taxicab industry and have been a driving force behind the rise of the "gig economy" that has spread to other sectors as well (see Kessler, 2018).

The ideologies surrounding and bred in Silicon Valley have also been disseminated widely through the products that the tech industry has created. And the sophisticated marketing campaigns that accompany technology products add to the aura that the products seem to take on. The yearly Apple keynote events, for example, in which the CEO of Apple unveils new products is covered by the press as an important media event.[6] Yet, as much as American consumers seem to imbue consumer technologies with specialness as the tech industry has grown and become more central to the U.S. economy, there has been an accompanying public awareness of the shortcomings within this industry. Silicon Valley's unwillingness or inability to hire people of color, especially Black people, has been both well documented in the media and mostly ignored by the tech industry (see McCorvey, 2015). Alice Marwick (2013) has provided another salient critique of the industry in her ethnographic account of Silicon Valley. She documents several instances of explicit and implicit sexism voiced by her informants. The widespread belief that the tech industry is a meritocracy, she argues, serves to doubly exclude women and people of color by first being run by men who do not notice their own bias and second by assuming that the lack of women and people of color in the tech world signifies their lack of drive or intelligence.

Silicon Valley, and the many places that seek to replicate it, spring from a specific type of business culture defined both by techno-utopianism and a reliance on neoliberal economic principles. Since its inception, Silicon

Valley start-up culture has aspired to "world-changing" impacts, whether in hardware or software pursuits, although what was meant by these lofty goals was often ill-defined. Still, because Silicon Valley has proven to be a site of innovation and profit generation, and because it has developed products that have become central to the habitus of U.S. culture, it has come to occupy a rarified space in the American imaginary. The entrepreneurs who populate the subindustry of faith-tech sometimes embrace and sometimes wrestle with the culture of Silicon Valley, but they believe in the potential of the technology industry and they have internalized many of its ideologies. This affects the kinds of products they create, the kinds of audiences they speak to, and the kinds of changes they (and their products) drive in evangelical culture.

MAKING A FAITH-BASED START-UP

Origins and Motivations

The oft-repeated fact that Apple began in Steve Jobs's garage in Los Altos, California, has become a parable demonstrating the possibilities of business greatness stemming from humble, hobbyist beginnings that connects to the ideological predispositions of a Silicon Valley–based understanding of meritocracy. The appearance of garages index this idea in commercials like the one produced by Xfinity in 2014 in which an Xfinity customer opens his garage and finds that his kids have created a tech start-up—with the help of their powerful internet connection of course—and are already experimenting with drone and hoverboard technology (Comcast/Xfinity, 2014). It has been reported that the CEO of Amazon, Jeff Bezos, moved into a house with a garage so that he could make the claim that Amazon began in a garage (see MacGillis, 2021, p. 30). At one point in my fieldwork the CEO of the company at which I was an intern abandoned the office to work in his garage with his CTO, presumably to get back to the roots of what tech innovation is about—the romantic feeling of the garage rebel.

Many of the founders of start-ups that I interviewed similarly highlighted their humble beginnings. CEOs of faith-tech start-ups, just like founders of any start-up, have a founding story that they repeat. The only difference is that some in the faith-tech world characterize their business ideas as having been guided by God or providence. One founder, who asked to remain anonymous, told me that the idea for what would ultimately become his faith-based app came to him when he was deep in meditative prayer. He said that after some setbacks in his tech career, he had asked God what his purpose was. As he described those formative moments:

> One thought kind of set in my mind and it was stillness around it, there was no other thought. It was no distractions but that thought, which was to enrich media lives by creating technology that empowers churches to provide greater impact to their communities. And I was sitting there and I felt like, "Oh this is it!" I felt assured, "I felt no doubt that that's what it is. That's my life goal!"

For many this sense of being called to their business is a real and deeply felt motivation for their endeavors. This connects with the common discourse in Silicon Valley about start-ups as agents of "disruption" or positive change. Most of the CEOs and founders that I spoke with did not put their founding story in spiritual terms but still retained this sense of calling.

Aaron Martin, one of the two designers behind the popular app Neu-Bible, explained the app's founding story: Martin and his business partner were tossing around ideas for collaboration while camping in Yosemite, and the idea for the Bible app felt right. He said it did not strike him as much different than a lot of start-ups whose goal is to create something that will be monetarily successful that will also have a social impact. He noted that in Silicon Valley, where he had worked for many years, it was common to have a secondary reason for your business—namely, that it would provide a social good. "I don't think it's necessarily a Christian idea," he told me, "but it felt more deeply rooted for us than some of the other ideas that we had toyed around with" (Martin, 2017).

The term "impactful" gets used a lot in Silicon Valley, and it connects to the ethic that the entrepreneurs I spoke to want to put out into the world. They want to make technology that can positively "impact" the world and often what is known in evangelical culture as "the Body of Christ"—referring to the global aggregate of all Christian believers. That their founding stories sometimes take on a spiritual dimension is not surprising, and in fact it is also indicative of the nearly spiritual understanding of the entrepreneur in tech culture. Often described as "visionaries," tech founders are elevated to near godlike figures in Silicon Valley. Alice Marwick, in her ethnographic account of tech culture, has noted that "the highest position on the status hierarchy is reserved for entrepreneurs" (2013, p. 80). Perhaps it is easy to mythologize the origins of a start-up given the cultural cachet these businesses and their founders have. Therefore faith-tech entrepreneurs do not flip the script when they describe their start-ups as driven by a spark of divine inspiration—they just add a Christian flair to it.

What these founding stories also point to is an understanding of the power of business and especially the technology business as an agent of change. The discourse of start-ups that sees technology companies as particularly important to social progress connects to the social imaginary of Silicon Valley's role in society, the idea that these businesses are all doing their part to "save the world," which many Christians in tech believe and have internalized. Being able to characterize activities such as coding, programming, and design as a life purpose indicates the power that tech holds as an agent of change in the American imagination. For Christians this possibility is tantalizing because for Christians working in tech, digital habitus, if guided in the right way, could make people better and more faithful. One venture capitalist I spoke to described the ways he could imagine technology transforming faith:

> You may not even have a mobile device, it will be your glasses or it will be contact lenses or it will be the shirt you are wearing and it will say, "Hey you haven't stood up all day long you've been sitting in front of this computer," "It's been thirty-six hours since you've actually seen your wife," or "You missed the ten-day streak of having bath time with your kid. Bad dad point." But technology should empower and help change behavior and get out of your way. All that meta-data, all that stuff is here, but making it useful and available and changing how you as an individual and as a collective group and as a society thinks.

This vision reads like science fiction, but the idea that technological products could improve the way that people live and even think is a common one in Silicon Valley discourse. It is particularly appealing to Christians, who want to help "redeem" or spiritualize the world. As this VC does, Christians interpret the world-changing promises coming out of Silicon Valley within a religious framework. If Silicon Valley entrepreneurs promise to "make the world a better place," then faith-tech entrepreneurs hope to make the world more faithful, more Christian place through their technological products.

Despite these goals, however, this type of business venture is precarious. Start-ups can easily fail, and many do. Many of the faith-based start-ups I interviewed had not received venture capital funding. Some had received funding from personal and business networks; sometimes even church networks, but most were self-funded. One CEO told me that money was often tight, sharing that once "we had no money in the bank and this friend was like 'I'll write you a $5,000 check,' so we just went up to midtown, picked up the check, put it in so we could clear our next payment." Yet these problems are common to start-ups of all kinds and also part of the romantic appeal of start-up culture.

Start-ups tend to be small. Some employ only two or three people who might not ever work in the same room together. Businesses that employ some transnational workers are common. At the start-up where I interned, for example, I never met the lead designer, who worked remotely as he traveled around the world. Conferencing apps like Zoom and Uberconference facilitate this kind of collaborative work over distance. As I interviewed founders, CEOs, and others in faith-tech, I sometimes met them in their home offices, sometimes in small, rented coworking spaces, and sometimes at the large tech companies that remained their day jobs. One start-up I visited was housed in a church—their church had offered to let them use the space. Typically start-ups don't advertise their smallness, as they worry it may be indicative to others (potential founders, business rivals) of a struggling operation. I once sent an email from a person that didn't exist for the CEO because he did not want it to seem like his business was just a one-man show (though it mostly was).

The smallness of these business environments allows for a lot of play to take place. The office takes on an intimacy as ideal Spotify playlists for workflow are discussed. Many faith-tech start-ups emulate the cultural spaces of the larger companies who inject a sense of playfulness into the work space. While visiting the Yahoo campus, for example, I glimpsed their doughnut wall—a large vertical wall covered with differently colored and flavored doughnuts. Facebook and Google famously have game rooms in their offices and hope to keep employees entertained probably for the purpose of keeping them on the work campus.[7] Small faith-tech start-ups also display and encourage this playful attitude toward work. The start-up where I was an intern had an old video game console installed in it, and the CEO asked me what I thought about converting part of the office into a "chill space."

Joanne McNeil (2019) has written about the culture of Silicon Valley with regard to the culture of Google:

> A stereotypical Google employee has perfect SATs, but loves April Fool's Day humor and a weekend with nature. Not all of its engineers are triathletes— some enjoy snow kiting and kayaking, too. There were office ski trips from the very beginning. Google is the summit of the Montessori-to-MIT pipeline for a person bright and logical who does not think Sergey Brin's toe sneakers are weird. (p. 23)

The style of tech businesses is often quirky but caters to a specific class habitus, as McNeill subtly points out. Tech businesses expect their employees to have a certain sense of humor, a certain set of hobbies, and none of these things are class-, race-, or gender-neutral. Why is this important? When faith-tech companies emulate the style and customs of the tech world, they adopt the biases that come with it. They hail a certain type of employee who is comfortable within a certain aesthetic framework. UX designers, coders, and CEOs come from or are assumed to understand this culture, which means that the products they create will reflect the habitus of that culture.

Fitting In

Although these start-ups emulate the aesthetics of tech businesses, use the same products, and speak the same language, even for those working in the central node of tech culture, they are often seen (and see themselves) as outsiders. For example, I met Aaron Martin at the cafeteria on Yahoo's Sunnyvale, California, campus in the spring of 2017. At that time Martin had worked at Yahoo for three years as a design director. In 2015, together with another friend who is employed at Facebook, Martin created Neu-

Bible. From the early days of beta testing the app, he received insulting comments. "We've had people say that Christianity is stupid," he told me, "and there's no way I would put this on my phone. I don't care how well designed your stuff is, I'm not helping you test this" (Martin, 2017). Others in the faith-tech space told similar stories. One faith-tech CEO who focuses on video production and works in Los Angeles told me that among his peers "if I told somebody that I was shooting porn, they would probably think it's cooler than if I said I'm shooting religion or faith." And the CEO of a large faith-based company in Washington, who started his career at Microsoft, said that he was used to the condescension he received from people. He told me: "I have a thousand episodes of cocktail parties of getting the funny look when you tell people you build Bible software."

For Aaron Martin, though, the flak he received from beta testers of NeuBible and others was par for the course and he accepted it in stride. Ultimately the app was covered by the tech press and received good reviews from well-regarded Silicon Valley publications. Although there is a challenge to branding a company as a Christian organization, Martin thinks that Christians are much too sensitive to not fitting in, especially those who did not start their careers in the tech industry. Often, he told me, Christians in other regions do not really understand Silicon Valley and assume that because it is a "less churched area," as a Christian he might face persecution for his faith. For Martin, however, that has not been the case. He said:

> I was at a start-up that was in Dallas so I've been through the South, that's where my Mom's family is from. There's an idea that I should have been persecuted for doing this, but I don't think that was the case. It would have been a great story like overcoming this persecution to make this great Bible app! But I think we knew it was going to be a hard business because we had multiple things we had to try and overcome the fact that it was a faith business that makes your market segment a little smaller so it's just another business challenge that we had to overcome. (Martin, 2017)

Martin gently pokes fun at what he views as Christian oversensitivity, and he thinks this is especially present in regions of the country where there are a lot of Christians. In these places the Christian echo chamber and the sense of the "establishment/ outsider paradox," to use George Marsden's (2006) term referring to the way that conservative Protestants often believe themselves to be an embattled minority even though they are represented and embedded in all aspects of the American power structure, contributes to the evangelical belief that powerful media industries like tech and Hollywood have values that are at odds with Christian culture and that all Christians working in these industries have to adopt a crusade-like mentality.

Others agreed with Martin that evangelicals have a tendency to overstate their outsider status. Silicon Valley is a religiously pluralistic place, but it is an ultimately agnostic culture. What matters most is what you produce, not what you believe. One technologist who works in a tech incubator in New York City said that he openly discusses his faith with his coworkers:

> We as Christians, I think, are called to be witnesses and, you know, say "Hey, what do you live for and what drives you?" And, you know, inevitably they're going to say "My love for my God and my love for my family, and really bringing renewal to all of the things that I touch and do." I think people respect that. I think people, when they think about "Oh, you take time to meditate? That's great! To not go insane? That's great."

Tech culture can be brutal. And this same technologist told me that he feels it is full of "type A alpha males" who see the industry as a zero-sum game. This technologist points to the other side of the problem of fitting in here. What does it mean for a Christian to fit into a cutthroat business culture? And what might be lost if one does?

For one CEO whom I spoke with, the culture of Silicon Valley and the relentless focus on success is one that he felt like he had to escape. Before starting his app, geared toward churches, he worked for years and was successful in Silicon Valley. Now, though, he still lives in the area, he does not attend the parties or social events as he once did, and he does not want to get sucked back into a culture he sees as toxic to his spiritual growth. "I know a lot of people in Silicon Valley are obsessed with success building the next big thing," he told me. "And closing the next big round by whatever means necessary. I think that's what I see more and more is, like, you lie, you cheat, you do whatever you want, but as long as you become successful, that's what people are going to talk about." This CEO believes that at one point he also became obsessed by the idea of making money and gaining prestige. It affected his faith and his marriage, he said. It was only after he let go of his place in the tech hierarchy that he felt like he could grow as a person. He put it to me bluntly: "I feel like more and more people in Silicon Valley need Jesus."

Prayer Hands Equals $100

With only two exceptions, all of the CEOs I spoke with ran for-profit businesses, and making money was a clear driver for their work, even when they also believed that their work was important or redemptive. Dean Sweetman told me that he had no problem garnering investment and interest for his venture Tithe.ly, an app that helps churches collect digital donations. The numbers supported it, as he explained: "Anyone who invests in a start-up is going to really look at the basics of the numbers. The market that we're involved in, you've got about $130 billion that are given to faith-based organizations in North America and 85 percent are still cash and check, so the enormity of the market is kind of a no-brainer for investors to back a company like ours" (Sweetman, 2017).[8] For Sweetman the religiosity of the audience means little to investors. Like any audience, their value to investors lies in their buying power.

When we spoke in 2017, Sweetman was pitching "emoji giving" to his church customers. Parishioners would be able to text an emoji that was associated with a level of tithing—for example, "prayer hands equals $100." He was also pioneering turning Facebook and Instagram likes into donations—so if a parishioner likes their church's Instagram photos a certain number of times in a month, that number would be multiplied by a dollar amount and determine their tithing for the period. Sweetman is using digital tools in an innovative way to appeal to the digital habitus of a young Christian audience, and his success in the faith-tech business proves what purveyors of Christian media have long known: Christians make up a big market in the United States. At the same time, Sweetman expressed a common sentiment among faith-tech entrepreneurs: the idea that their business, because it is run by Christians, is different than other start-ups. "I think to say that we're just a company that's about trying to get to the biggest market share, to try and sell this company so we can all make money, that's really the furthest thing from our intention. Our business is to serve the body of Christ with great technology" (Sweetman, 2017).

Influenced by his church background—Sweetman was a pastor before he became a businessman—he noted clear differences between the culture of church and the culture of business:

> In the church world things are very black and white, and you've got a clear guide on how to live, and if you're in adherence to the Bible, you have a guide on how to live life and treat people and so on. Business is almost the opposite, it's anything goes. It's all just about money. So if your motive is just money, you are going to cross lines all day long, but if your motive is to do good and help people with good products, then I think you can live within those boundaries and ultimately add value to not just your business but to the community that you serve. (Sweetman, 2017)

Sweetman says that money is not the evil, but the obsession around money in the tech business can be. His sentiment gets to the root of the problem that faith-tech entrepreneurs face: they want to make money, many even expressed that they had started their business because they thought it was a good business opportunity. Maybe, some even thought, they would found the next "unicorn"—a term for that rare start-up that earns a billion-dollar valuation. Yet these hopes for financial success had to be negotiated with Christian values.

Sweetman's understanding of business is illustrative of the bind that many Christian tech CEOs find themselves in. They are businesspeople who run for-profit businesses, but they also understand that their faith makes them different. They have to actively construct their companies and their own images in such a way as to represent what they believe to be Christian values, and these are not always compatible with the values that dominate Silicon Valley start-ups. Some faith-tech CEOs shrugged off this problem as just one of optics—because they are explicitly running Christian companies and targeting Christians, they have to be sure that their branding appeals to their demographic.

But many noted that this reliance on a religious audience was sometimes a burden. One CEO who has been in the faith-tech sector since he started his company in 1991, noted that "people call us to a higher standard. I get a different set of creative insults from customers who are angry. Because people want to use the spiritual stick to beat you with or something. Something happens, like the product crashes, [and] 'Well,

I expected higher quality from a Christian company!'" This CEO used a comical example of a petty complaint, but he voiced a common theme that came up in my interviews. When a company brands itself as Christian, it has to make sure that its public face is Christian *enough* or it risks facing backlash.

For another entrepreneur working in the VR scene, whom I call Trevor, this reality has been an annoyance. He told me that people are constantly trying to judge how religious or spiritual he is because he runs a company aimed at a Christian audience. And although Trevor is a committed Christian, he says (while gesturing to his heavily tattooed forearm): "I'm not exactly poster boy Christian here." He does not want people to judge him but instead to judge his company, and he uses the metaphor of parenting to talk about how he runs the business ("The company is the baby I'm growing up"). It is important for Trevor to imbue his Christian values into the business, but as the metaphor implies, he wants there to be a separation between his persona and the character of the business. This entrepreneur believes a separation is particularly important in the faith-tech sector, where there is a different set of expectations from faithful customers and investors.

Thus there is always a negotiation between running a business and being a faithful person. Although there is money to be made in the faith-tech sector, money-making cannot be the only goal of faith-based start-ups. The ways in which faith-tech entrepreneurs interpret this balance sets their subindustry apart from that of Silicon Valley and reveals that the drive to create a faith-based start-up is not just about money—it is about serving and sometimes changing or updating evangelical culture. Or, in many cases, it is about using consumer technologies to Christianize American culture.

Redemptive Entrepreneurship

I spoke with two Christian founders of an app company that does not label itself as Christian. For them it was a business decision first: Why would you limit your market share? But it was also a reflection of their understanding of the role of Christians in culture. One of the founders told me that when he sees companies brand themselves as Christian, he thinks:

Just do what you're doing. Do the thing you are doing and do it for every-one. I think U2 does a good job of that, but Jesus does a good job of that. Be excellent in what you do and offer really great service that meets people's needs and do it excellently, but there's this unneeded step of being like "... and we're Christians!" If you do that, you do this subculture identification and you kind of win this group of people, but you also ostracize yourself from a lot of other people because you are doing this for the Christians and it's like, why not just do it for the world and do it really well?

This entrepreneur does not want to specifically focus on a Christian audi-ence for his app not only because he does not think it is a good *business* decision but because he does not think it is a good *Christian* decision. A good product is more likely to have a positive impact on the world than a specifically Christian one, in his view. Christian businessmen might see their businesses as a chance to connect to people, in the way that a song by U2 might connect to listeners, and they believe that the values that they imbue into their businesses might translate to those people and influence them in some way. He references Jesus, who did not "preach to the choir" but brought his message to everyone. This should be the basis of a faith-tech business, not "subculture identification," according to this entrepreneur.

A venture capital company called Praxis, whose goal is to fund and incubate companies with Christian missions, is illustrative here. By actively promoting, training, and supporting Christian entrepreneurs, Praxis hopes to inject business culture with Christian values—although the businesses that they incubate do not have to only cater to Christians and they do not have to explicitly brand themselves as Christian busi-nesses. Praxis takes a "missional" approach to thinking of business, which means they see work (and particularly entrepreneurship) as a place in which Christians might influence society. As with many of the start-ups I have discussed so far, this vision comes from an understanding of the power of businesses in American culture.

The founder of Praxis, Dave Blanchard (2014b), has written: "Only as we step back and explore what seems unthinkable can we truly begin to imagine creating ventures that might shift and shape culture. We might think of this as the cultural renewal analogy to Clay Christensen's ubiqui-tous *disruption* terminology. With our short lives and Western privilege, what if entrepreneurs aimed to be cultural pioneers and explorers instead of moguls?" Business is the most powerful agent of change in culture, in

Blanchard's view, and thus this is where Christians need to be. Christians cannot be normal businessmen, though; they must serve a higher calling. Although Blanchard privileges business, he sees it as a sector of society that needs to be "redeemed" or imbued with Christian values. "It is hard work maintaining integrity and a right motive in the world of Mammon," he wrote, "which is consistently and ruthlessly recruiting us to the lesser, decadent pleasures of power, prestige, and possessions, each of which can be instead used winsomely to benefit others to His glory" (Blanchard, 2014a). The goal, for Blanchard, is not to fund and incubate a crop of companies that are simply going to make money (although of course that is part of it). Rather, the goal is to incubate companies that Praxis believes will spread Christian values by creating a work culture that does not make profit the ultimate goal but instead takes glorifying God as its mission. Like the CEO who feels he needs to avoid Silicon Valley parties because they are toxic to his faith, Blanchard criticizes a culture obsessed with "lesser pleasures." Yet, because this culture is influential, Blanchard implores Christian people to brave it.

Praxis has funded various companies, most of which do not have an explicitly Christian mission. The Giving Keys is one such venture. This business employs homeless people in downtown Los Angeles to make necklaces stamped with simple, positive messages that say, for example, "love." Although the company does not advertise its religious background, the business model does not emphasize profit above all else. It also seeks to provide a public good: in this case, to empower homeless people. Praxis also funds Lasting, a marriage counseling app, and Cladwell, a start-up that uses artificial intelligence to help people curate their clothes and buy fewer products. Although none of these companies are explicitly Christian, in keeping with the philosophy of Praxis, the businesses hope to imbue their work and by extension American culture with Christian values. Praxis calls this strategy "redemptive entrepreneurship."

In 2017, at Santa Monica a coffee shop run by a local church group, I spoke with a venture capitalist, whom I call Randall, who has worked alongside Praxis about Praxis's goals. Randall reiterated Dave Blanchard's vision of the role that businesspeople can play in cultural change, saying: "Ultimately we want to influence culture. So that's absolutely our goal and we are very intentional. If you were to ask yourself back in 1950, how does the parachurch or the church influence culture, we would have one set of answers. If you were to ask that question today, we think that entrepreneurship would be the key through which to influence culture." For Randall there is a biblical precedent to this: the Apostle Paul, he explained, evangelized through his tent-making. "You know there was a reason why Paul was in the Agora in the marketplace," Randall told me, "because that's where things happen." Entrepreneurship is where Praxis believes that culture happens, and so, theologically, it is a place where Christians need to be and need to be thinking about their role in culture.

Randall has worked for many years in venture capital and has focused on Silicon Valley. Perhaps because of this he is more gimlet-eyed than many about the role of Silicon Valley as a great savior. "When you work in tech you are around a lot of positive—oh this is great stuff! This is exciting! And yet you say to yourself, you know, that's true and we are very fortunate to be part of it, but you know is humankind really different today? Is humankind really interested in more human flourishing than it was twenty years ago?" He makes the point that although Silicon Valley has often promised to solve the world's problems, these tech companies have done very little to deliver on those promises. For Randall and Praxis, one way that business might begin to do some of the work of helping "human flourishing" is by starting with Christian values. For Randall this happens on both micro- and macro-levels. It happens when Christians are in the workplace, he says, providing an example for their coworkers through their behavior. It happens on the macro-level when businesspeople make decisions based on Christian values rather than bottom-line values. The philosophy of "redemptive entrepreneurship" trusts in the premise that business, and especially new start-ups, can save the world but only if Christians can first save business culture.

Redeeming the Internet

Praxis and others in the faith-tech space see their mission as using businesses and business culture to promote Christian values, but the idea of redemptive entrepreneurship also extends to those products that hope to redeem digital habitus. Although evangelical Christians are generally excited about the possibilities new media technologies can bring, they understand that internet use also has what they view as a dark side. Christians often characterize the wide availability of internet porn as a great evil that has the potential to ruin families and destroy the fragile psyches of young Christians.

As the landing page for Covenant Eyes' (2020) site states over a picture of a child whose face is lit by the eerie glow of a laptop's screen: "Viewing porn releases powerful, mood-altering chemicals that literally rewire your mind, until you crave it more than authentic human connection." As they put it: "Porn is the ultimate villain" (Covenant Eyes, 2020). Covenant Eyes is an app that pairs each user with a partner who gets a copy of their partner's browsing history every month in order to hold them accountable and keep them away from internet porn. Covenant Eyes uses technology tools to fight the perceived ills of technology, and they port a strategy common in church small groups where "accountability" to other Christians is meant to help Christians through their "walk."

XXXChurch.com (Triple X Church) uses the same accountability strategy on their app, and they also do in-person outreach with the porn industry. I spoke with an employee who works on the app and sees her own story as proof-positive of the power of this model. A former porn actress, she discovered Triple X Church at a porn convention where she was signing autographs for fans. The founder of Triple X, Craig Gross, gave her a Bible and connected her to his ministry. Gross had been a youth pastor who felt the app would help the teens he saw struggling with porn addiction. After meeting Gross, she told me, her life was "radically transformed." In addition to working for Triple X Church with her husband, she is now a minister at an evangelical megachurch in San Diego, California. She also engages in evangelical outreach with the sex industry through Strip Church, an organization that proselytizes to women working as exotic dancers.

The stories of Covenant Eyes and Triple X Church illustrate how some Christian technologists believe they can use consumer media products to redeem internet use, another aspect of the strategy of redemptive entrepreneurship.

Shortcomings

The technology industry centered in Silicon Valley has exported potent myths about the power of consumer technology. For example, the industry has created new cultural ideologies of "disruption" as a business goal, and it has tended to characterize its products as "revolutionary"—recall the famous "1984" commercial for Apple computers, which had an athletic rebel throwing a sledgehammer through a projection of Big Brother on a movie screen while being chased by gray-clad figures. When Christian entrepreneurs internalize the ideologies of disruption and revolution, and filter them through a Christian lens, they locate redemptive hope in the production of consumer technologies.

Yet despite their sometimes lofty goals, the faith-tech industry displays many of the same problems endemic to the larger tech industry. There was nearly no diversity in the start-ups I researched. Nearly all of the people that I spoke with for this book were men, and nearly all were white or of East or South Asian descent. Although this research was not exhaustive, it was nonetheless clear that the world of faith-tech is primarily white and male-dominated, just as the larger tech industry is. While the faith-tech industry produces technologies they hope can reach the world on a global scale, they have mostly been created within this white, male-dominated cultural context. Although gender politics in evangelical culture are complicated—as I explore in chapter 4—men still occupy most leadership roles. Many of the men I spoke with talked about supporting their wives and families as part of the anxiety that surrounded the success or failure of their businesses. Many had wives who did not work outside of the home, which is common in evangelical culture.

In terms of racial diversity, faith-tech reflects both the problematic messaging of Silicon Valley culture and of evangelical culture. On the one hand, the technology industry purports to be a meritocracy; yet on the other hand, it consistently fails to hire candidates of color. As in evangelical culture, racism is seen as a faraway problem. The understanding is that individuals should not be racist. As the influential evangelical Rick Warren (2014) explained in a Tweet: "Racism is a SIN problem, not a SKIN problem." Beyond that, however, there are very few structural understandings of racial exclusion put forth by white evangelicals, just as there are few attempts to meaningfully address diversity in the tech industry. The unspoken consensus is that the market will take care of the problem and that good people will rise to the top. This sentiment is so widely shared among white Christians that it obscures the structural and cultural factors that make racial parity an impossibility.

The strategy of "redemptive entrepreneurship" relies on the idea that if Christians are in business, business will be better, but apart from imbuing values and potentially evangelizing or providing an example to coworkers, there is no coherent strategy to change the tech industry for the better. Thus, while faith-tech companies hope to both change the tech industry and change the world through tech by imbuing digital habitus with Christian values, the industry has in a sense remade the tech industry in their own image and has duplicated many of its shortcomings.

CONCLUSION

In the spring of 2017, the enthusiastic entrepreneur I met at a New York incubator, Jack, tweeted that his start-up would be suspending operations. Like many start-ups, it had reached the end of its run. Two months later Jack put up a vlog where he explained that his start-up had been bought. I caught up with him and he said that ultimately he was really happy with the way things had gone with his app. No, it had not been a unicorn and he was not going to get to retire, but that was the business. You start something, see what happens, and then hopefully get a chance to start something else and the fact that he had been bought by a Christian company with a lot of resources was a good fit. Jack had been CEO, CTO, and CFO of his company for two-and-a-half years and he was ready to move on.

The business of tech start-ups is a volatile one. It moves quickly and sometimes relentlessly. I met several entrepreneurs whose initial hopes had not panned out and some who seemed genuinely jaded by the tech industry. Despite the fact that all of the people I spoke to who had started faith-based companies believed that they were doing important work, there was certainly a realism and a general malaise in the air. Yet Christians remain in the industry, and new faith-based companies pop up regularly because Christians in faith-tech see business (and particularly the tech sector) as a way to gain influence in American culture more broadly. Not unlike Christian political activists who have supported "values candidates" since the 1970s, these evangelical businesspeople theorize their place in technology as a means to "influence culture," in Praxis's terms. What this reveals is the power that business, and especially the myth of the tech start-up, has in the evangelical imagination. The tech start-up is seen as a place where real "disruption" might happen and where Christian values might break through into mainstream culture by becoming integrated into digital habitus. In this way the discourses and practices of these evangelical entrepreneurs follow the logic of neoliberalism, the idea that free-market businesses pave the way to a better society. Government policies are not seen as effective means of influencing or changing society, but entrepreneurship is. And the most powerful type of entrepreneurship and the most lauded in the American imagination in this moment is that of the tech start-up. The products that this industry has created add to the assemblage of digital tools and products in Christian culture. In their implicit embrace of technology and the technology industry, these products influence the Christian imaginary of tech. Each of these apps and websites is a piece of digital habitus, and the fact that they continue to proliferate is further evidence of the seductive power of digital habitus as a modernizing agent in evangelical culture.

By seeking to infuse the technology industry with Christian values, and by creating products that they hope will aid spiritual development, faith-tech businesspeople see digital habitus as a dual opportunity: to make money and perhaps to "save the world." As in the church world, they embrace the corporatist sloganeering and the strategies of Silicon Valley, and they see technology as a means to remain relevant in contemporary American culture. But, as is discussed in chapter 3, their attitudes also display problematic blind spots primarily by resting on the assumption that technology is culturally neutral. As evangelicals attempt to take technological innovations into the missions field, they run into problems with the corporatist leanings of the faith-tech sector and are forced to rethink how technology and the technology industry is culturally and racially coded.

3 Media Missions

On the second floor of a small stucco office building in Torrance, California, John Edmiston and his partner make raspberry pis—tiny single-board computers equipped with Wi-Fi and Bluetooth—to be shipped to remote areas of the globe. Edmiston (2017) explains that the device has to be "ruggedized" so that "it can be dropped, it can be in mud, it can be out there. It can be in heat, it can be in dust, it can be in the tropics." This is the latest project of Cybermissions, Edmiston's nonprofit. Edmiston hopes these "Pastor Boxes" will offer seminary-level training for indigenous Christian leaders in locations where mobile phones are plentiful but internet and cellular network connections are scarce. This project fits into the stated goal of Cybermissions, which is "using computers and the internet to facilitate the Great Commission" (Cybermissions, n.d.). Edmiston is part of a network of Christian nonprofits and individuals who believe that technological products when conceived in the framework of Christian missions work can be used to evangelize the world or fulfill what evangelicals call "The Great Commission."

Evangelicals trace this missionary imperative to the Bible, specifically to the verses in the Book of Matthew that describe Jesus telling his followers: "Therefore, go and make disciples of all the nations, baptizing them in the name of the Father and the Son and the Holy Spirit" (Matthew 28:18–20). Another verse in the Book of Revelation sets the stakes for this missionary work when it describes a vision of the end of Christian history: "Then I saw another angel flying in midair, and he had the eternal gospel to proclaim to those who live on the earth—to every nation, tribe, language and people" (Revelation 14, p. 6). In my fieldwork and interviews it was not uncommon to hear evangelicals explain that they believed that they might be part of the end times and that spreading the gospel to every nation, tribe, and tongue was an agentive force in fulfilling this prophecy. Many people expressed that fulfilling the Great Commission was the most important task set out for Christian believers. And this imperative structures how evangelicals see media use, especially new technologies in missionary work.

This chapter documents the ways Christian missionary organizations view digital tools. I begin with a focus on the most visible proponents of digital missionary work, those involved in what is known as "mobile ministry." This is a world in which self-proclaimed "tech geeks" come up with workarounds to get Bibles into underground churches in China by practicing what is known as "SD card evangelism." In this world, programmers rejoice when their apps are opened in Saudi Arabia—a country notorious for its rigid antiproselytizing laws. Coders construct secret Bible apps that cloak their religious content. It is a world in which the complex and often problematic history of Christian missionary work is in some ways being challenged and in some ways being continued by a new set of digital utopians. These techies often echo the rhetoric of those in Silicon Valley who want to "save the world" by spreading technological products and digital habitus.

Yet, unlike the other spaces this book explores—faith-tech entrepreneurs, networked and online churches, and Christian social media users and podcasters—this is a space that challenges the norms of corporatist technology in favor of the early visionaries of the web and especially the discourses of the Free and Open Source Movement (FOSS). This is significant because it allows for a Christian critique of digital habitus as it relates to the prominence of corporate-produced consumer technologies. Missionaries using digital tools to communicate and proselytize have had to ask questions of technology that churches and faith-based start-ups, whose audience tends to be the white American evangelicals, have not. Because of this, they understand technology in different ways. Digital missionaries have taken up the digital utopianism from the internet's early boosters, but as they have translated their enthusiasm into cultural products and missionary strategies, they have faced a host of problems that reveal the cultural biases of technology. Although individual cases are incredibly complex, scholars who have studied the long history of evangelical missions projects agree: as American evangelicals engage with global Christianity, they change. This chapter tells the story of how evangelicals are using technological tools to reimagine the project of missions and how this project is changing the way that evangelicals see the world.

THE MISSIONARY TRADITION IN CHRISTIANITY MEETS DIGITAL EVANGELISM

In November 2018 a young American missionary named John Chau was killed as he attempted to contact a tribe living on a tiny island in the Bay of Bengal. When news of Chau's gruesome death by arrow reached social media, he was widely ridiculed for his hubris and cultural insensitivity. People took to social media to lambast him and the missionary impulse that had led Chau to believe that he needed to evangelize to this "unreached people group." He was referred to as "a stupid colonizer," and one Twitter user said Chau was "closer to a murderer than a martyr."

Even among evangelicals, Chau's path was controversial. To many, Chau represented the worst of what missions work can be: arrogant, over-eager, and underprepared. But his death also served as a reminder that American Christians still see missions work as central to their ethos and worthy of the ultimate sacrifice. Ed Stetzer, a prominent evangelical pastor and author, reflected about Chau's death in the *Washington Post*. Stetzer agreed that Chau's approach was problematic, but he offered a defense of Chau's missionary zeal: "In today's world, the missionary mind-set itself is a modern-day heresy. However, it is still the teaching of Jesus and cannot be erased from the pages of the Bible" (Stetzer, 2018). The history of Christian missionary work is littered with martyrs like Chau—for example, the American missionary John Birch, whose name was immortalized in the right-wing think tank The John Birch society, was killed in China in 1945. Missionaries are held up as the most selfless and Christlike people in evangelical circles. And because they often give up Western comforts in order to spread the gospel to other parts of the globe, they are seen as particularly pious and devoted.

Because of the prominence of the missionary project in evangelical culture, short-term missions work, which sees people going for weeks at a time to "serve" communities around the world, has been folded into church curriculum for young people throughout American Christianity. And although the ubiquitous "mission trip photo"—typically a photo of a white teenager in an "exotic" setting, posted on social media—is often criticized as a problematic emblem of American Christianity's seeming insistence an imperialist mind-set rooted in a racist worldview, short-term missions work continues to grow, and myriad parachurch organizations devoted to organizing these missions trips have arisen in the past twenty years.[1] Kathryn Joyce (2013) has also made the case that this emphasis on "voluntourism" in far-flung locations fuels the "orphan theology" movement in evangelical culture that has spawned hundreds of parachurch organizations devoted to international adoptions. This has led to what Joyce calls "orphan fever" in evangelicalism—the near contagious spread of international, often interracial adoption in church communities. Although earnest believers may adopt with good intentions, the Christian international adoption industry is littered with stories of exploitation and fraud.

These highly visible expressions of American Christianity's problematic devotion to international outreach—the death of John Chau, the mission trip photo, the complicated geopolitical and racial dynamics of international adoption—are signifiers of what is sometimes characterized as "the white savior complex." This mind-set comes from the project of European colonization and was laid out by Rudyard Kipling in his infamous poem "The White Man's Burden" (1899). In the opening lines of the poem Kipling marries the white supremacy of colonization with the missionary project:

> Take up the White Man's burden—
> Send forth the best ye breed—
> Go send your sons to exile
> To serve your captives' need
> To wait in heavy harness

On fluttered folk and wild—
Your new-caught, sullen peoples,
Half devil and half child

The "white man's burden" refers to the responsibility to proselytize and educate those who are uncivilized, who are "half devil and half child." This poem reflects the problematic and racist project of colonization, which has been driven by an understanding of both racial and religious superiority.

But, if on the surface the modern missionary movement in America still seems to reflect this mind-set, the story is also deeper and more complex. The history of missionary work, or missiology as it is known in seminaries and divinity schools, is vast and comprised of myriad theological, historical, and ethnographic monographs. Christian scholars see the paradigm for missionaries traced out in the Book of Acts. Christianity from this time, as Lamin Sanneh (2008) has written, was a religion that relied on an understanding of the centrality of individual believers, rather than on a homeland or a sacred, revealed language, as, for example, Islam does. Because of this, Christianity was conceived of as a mobile faith that could move throughout the world and express itself in any cultural context. These theological underpinnings opened Christianity to the possibility of linguistic and cultural translation.

As it moved throughout territory and history, Christianity became cemented as the religion of Europe. During the Age of Colonization, missionary work was folded into and informed the imperial project. Christian missionaries were cast as civilizing forces. As Sanneh explains, many believe that in this era "Christianity was already so firmly anchored in the Enlightenment milieu of its origins in the modern West that in whatever forms it emerged in the rest of the world it was bound to sow the seeds of its formative Western character" (2008, p. 217). But Sanneh's research complicates this view of missionary work as a one-way process, inextricable from the history of Western imperialism. Jay Riley Case (2012) has similarly pushed back on this historical narrative to argue that

the missionary encounter did not simply encompass imposition and resistance, as many scholars have painted it, or simple proclamation and acceptance, as many evangelicals have described it. New movements of world Christianity emerged from a complex process of engagement in which local

Christians selectively took resources brought by missionaries and adapted them to their own contexts." (p. 7).

Case makes the point that missionaries have not proven particularly good at changing or "civilizing" Indigenous cultures and have been just as likely to adapt their theological and religious frameworks to fit Indigenous cultures. He uses the Mukti Revival as an example. In 1905 a group of American Christian missionaries at a girls' school in India heard about a young woman who had had visions of fire that she interpreted as visitations from the Holy Spirit. As Case recounts, these missionaries, excited by the idea of these supernatural visitations, actively encouraged them, sought out their own visions, and ultimately brought back a new form of Pentecostal Christianity to North America that later became a primary religious form in South America.

As this dizzying story reveals, missionary work was never a one-way movement. It has always been a complicated exchange of cultures and theologies that has influenced the shape of global Christianity. Case has also noted that missionaries who returned from the field in the late nineteenth and early twentieth century were among the only voices challenging accepted beliefs in white supremacy and civilization. For Case (2012), "world Christianity did more than academic theories of human difference to undermine racism in nineteenth century America. More often than not, the academic trends of that era intensified racist thinking" (p. 10).

Melani McAlister has similarly illuminated the complex push and pull between American evangelical institutions and global Christianity. Whether the issue has been the Iraq War or the Draconian antihomosexuality laws passed in Sudan, the popular media narrative in the United States has been that American Christians have outsized influence on the Christian populations of other countries. But McAlister highlights the myriad ideological and doctrinal struggles that have engaged American and global Christianity. For example, in the 1960s, while Christians were evangelizing in what is now called the Democratic Republic of Congo, the international coverage of Civil Rights struggles in the United States tarnished the reputation of missionaries abroad and prompted conversations about race among white evangelicals in America.[2]

McAlister (2018) notes: "However fraught, the cultural interactions between missionaries and Congolese were never just morality plays in which Westerners tried to impose their cultures while locals resisted (or acquiesced)" (p. 35). McAlister traces the global understanding of faith that American evangelical Christians have, and she explores how this understanding sometimes challenges the popular or political narratives of the moment. "American evangelicals might see their country as a force for good in the world," she writes, "but they also sometimes insisted that US foreign policy had gone dangerously wrong and needed to reform" (McAlister, 2018, p. 3). Christians understand themselves as part of a "global Body of Christ"— a phrase I often heard in my fieldwork—and one that has set the stakes for American missionary work in the past and currently sets the stakes for technology in the missions project.

Evangelicals have shifted their strategies at many points to react to shifting geopolitical realities. Contemporary missiology has identified what it calls "unreached people groups," a term dating to the 1974 missiological conference in Lausanne.[3] The idea that groups of people might be targeted by more specific metrics than their nationality or race shifted what missions work meant. Rather than identifying countries to send missionaries, Christian organizations now identify "people groups" by their ethnolinguistic identifiers. By dividing the globe into ethnolinguistic communities that sometimes transcended national borders, the Christian missions movement again reframed the missionary project. The emphasis was placed on translation of the Bible into thousands of dialects representing "unreached people groups" or "the unreached"—terms referring to those people who had not read the Bible in their native tongue or had not come into close contact with Christianity.

Later, in 1989, missions organizations would identify what evangelicals call the "10/40 window" front and center in an attempt to focus on the Middle East as a site for missions work. The 10/40 window, also known as the "resistant belt," refers to the area at 10 to 40 degrees north latitude and the areas that surround this point. This area comprises much of North Africa and the Middle East, places where Islam is typically the dominant religion. Evangelical missionary organizations in the 1990s and early 2000s began to focus on converting people of Muslim faith in majority-Muslim countries, which followed from evangelical theories of the "resistant belt." This became a problem when the Bush administration agreed to allow missionaries into combat zones in Iraq, a move that many foreign policy experts saw as a dangerous conflation of military and religious goals that signaled to many in the Arab world that the United States was fighting a war against Islam.[4] And thus American missionary work, despite what many ecumenicists have hoped, remained closely and tumultuously married to U.S. geopolitical power in the twenty-first century.

The imaginary of the Resistant Belt has also led evangelicals to understand their work as constrained by governments hostile to their in-person missions. Here is where media plays an important role. When evangelizing cannot take place in person due to antiproselytizing laws in places like Saudi Arabia, evangelizing can still take place—albeit illegally—through film, through printed tracts, through tapes, or CDs. Media technologies transmit the message of the gospel where people cannot go. And because of this, media has become central to understanding how to preach to those in "the resistant belt." Digital habitus is global, evangelicals assume, and so missions can happen through technology, through media. This is why tech-minded evangelicals I spoke with tended to concentrate on evangelizing in Saudi Arabia as an ultimate goal.

Just as the history of Christian missions is linked in complicated ways to a history of imperialism, white supremacy, and American exceptionalism, it also cannot be disentangled with the history of media use and technological innovation in Christian culture. Evangelizing through printed tracts, through radio, through television, and then satellite and cable has historically been framed in terms of reaching the unreached and spreading the gospel.[5] The movement of digitally minded missions organizations I look at here has had many antecedents. For example, in the world of Christian filmmaking, the famous American evangelist Billy Graham started Worldwide Pictures in 1953, which produced films meant to be shown all over the world as a way to convert people to Christianity. The idea that film is a particularly useful and potent mode of evangelical expression is still commonly voiced today. For instance, the popular project "The Jesus Film" explains on its website: "We believe film is the most dynamic way to hear and see the greatest story ever lived—so we are driven to bring Christ-centered video to the ends of the earth" (Jesus Film, 2020). Films can be produced in the West and distributed throughout the world, and they have the advantage of being easily transportable.

These projects also reflect a belief that film is a particularly powerful and persuasive medium.

The missions-focused evangelicals behind these media projects seem to believe in the hypodermic needle theory of communication—the idea that media can have direct, immediate effects on viewers, an idea that media studies as a discipline has been bucking against since the 1940s. For Christians this understanding of media easily fits into a spiritual framework. Evangelicals believe that the gospel is holy in itself, the gospel—guided by the Holy Spirit—can transform people. Given that the history of Protestantism relies on the doctrine of *sola scriptura,* the idea that a good Christian life and Christian salvation is achievable through reading the Bible, any medium that can get the Bible or the story of Christ into the hands of people who cannot otherwise access it is powerful. Evangelicals, then, are simply spreading what is colloquially termed "The Good News" and using any media they can to do so. In the contemporary U.S. consumer technology saturated landscape, this tradition has been amplified by those promises coming out of Silicon Valley that claim that technology can be a way to positively impact the globe.

For example, Nicholas Negroponte, head of the MIT Media Lab, conceived of the One Laptop Per Child project in which cheap computers would be given to children in underserved areas of the globe. The project's assumption was that the technology itself could bring about positive societal change just by being present in people's lives. Although the project did little to alleviate poverty or raise literacies levels,[6] projects that employ rhetoric like this are common in American technological and business circles. Morgan Ames (2019) has charted how the conception and implementation of the One Laptop Per Child Program imbued the technology itself—cheap, sturdy, portable computers—with charisma by connecting them with popular ideologies of computing and nostalgia for the "technically precocious boy" that resonated with technologists but were not necessarily backed by solid research.

Understandings of charismatic technology borne in the tech industry, when met with the long-standing evangelical enthusiasm for using media in global missionary work, inform each other rather naturally. Furthermore, the digital habitus in American evangelicalism is influencing new ways of thinking about the mechanics and ethics of mis-

sionary work among an influential network of digitally minded missionaries, inspired by the spread of digital habitus around the globe. But, as
the complicated history of international missions work shows, American
Christianity is never unchanged when it reaches into other cultures. As
Christians use digital tools to proselytize across the globe, they run into
new understandings of technology that circulate in evangelical culture
and challenge the widespread acceptance of digital habitus that has been
cemented in the American imaginary since the turn of the twenty-first
century.

THE NEW ROMAN ROADS

Given that technological innovation has been a part of missions work for at
least one hundred years, it is not surprising that missions-focused Christians and organizations would see the new media revolution as an opportunity to find new ways to proselytize. The most visible and institutionalized form that this has taken is the movement known as "mobile ministry."
"Mobile ministry" refers to the idea that mobile phones—not necessarily
smartphones—and the suite of technological products that come along
with them (such as e-readers, mobile apps, sim cards, SD cards, and video
players) provide a new way for Christians to get their message out to a
global audience that is increasingly adopting digital habitus. Some of the
people that mobile ministry hopes to connect with could be considered
"unreached people" in the Christian sense of the term—they have never
been contacted by Christian missionaries, but they have nevertheless been
reached by mobile technology.

The Mobile Ministry Forum, started in 2010, represents a network of
about 125 missionary organizations. Some of these organizations are large,
like the International Missions Board, which is the missions organization
that serves the Southern Baptist Convention (SBC), the largest evangelical denomination. Others are small, like John Edmiston's Cybermissions,
a two-man show. All of these organizations are devoted to the idea that
new media, especially mobile tools, should be the focus of Christian innovation and experimentation. As explained on their website, "The Mobile
Ministry Forum is a network of missional innovators fostering a mobile

ministry movement so that every unreached person will have a chance to encounter, experience and grow in Christ through their personal mobile device" (Mobile Ministry Forum, 2017).

Because of the growth and ubiquity of mobile technology, the people involved in the Mobile Ministry Forum believe that all people on the globe will be able to hear the gospel in their lifetime. In 2010 the forum set a goal for fulfilling the Great Commission in ten years, by 2020—although it was understood among them that this was a purely aspirational, likely unattainable hope. When I asked Keith Williams, one of the leaders of the Mobile Ministry Forum, about this goal in late 2017, he said: "That was ten years from when we first got together. And boy, I'm just watching something 100 million more people just got on the internet in India last year. The dream becomes more achievable even if we are not necessarily moving the ball forward as fast as we would like. God is moving that ball forward is what I would say" (Williams, 2017). As digital habitus becomes normalized in more and more cultures, Williams believes, so does the opportunity to spread the gospel through mobile-connected devices. The vision of the Mobile Ministry Forum and the movement known as mobile ministry shares a lot in common with those techno-enthusiasts who believe that closing the so-called digital divide is the most important step in curing global inequity. Again, it is reminiscent of Nicholas Negroponte's ill-fated One Laptop Per Child Program. Both assumptions rest on the idea that the technology itself is charismatic or even emancipatory and once every person has access to it, the world will fundamentally change. Of course, for mobile ministry the message is just as crucial as the medium, and getting the proper message transmitted through the proper medium is the central goal of the Mobile Ministry Forum.

Williams told me that his revelation about the power of mobile technology happened while he was a missionary in the field. He had gone to the Arab world (he declined to specify which country) armed with cassette tapes meant to transmit culturally relevant biblical messages, because research had shown that many in the area drove old trucks with built-in tape players. When Williams got to the region, however, he found that the cassette players were mostly inoperable. "We started seeing that they were putting up their tents where they could get cell phone reception instead of where they could get water and that was kind of an Aha! moment"

(Williams, 2017). This led him to think deeper about the role that mobile technologies have to play in the Christian mission to evangelize the globe. Williams used a metaphor that I heard echoed by two other interviewees, that of the Roman roads. He said:

> We compare the Roman roads that Jesus and his disciplines walked down and we say, God knew the perfect time to create a place where Jesus would come in right at the time when these Roman roads were connecting the entire world and that would just be this avenue that could be used for the spread of his Kingdom and we look at in the same way right now with mobile telephony. (Williams, 2017)

Williams sees the world growing ever-more connected through mobile technology, and it is through this technology that Christians can spread the gospel, just as the Roman roads facilitated the spread of Christianity in biblical times. He expressed a sentiment that I heard a lot in my fieldwork, the sense that technology has been given by God and that Christians should use it, or as they put it, "steward" it in a way that expresses their message. This idea syncs well with rhetoric endemic to Silicon Valley. Like many technologists, these Christians believe in a determinist vision for the New Roman Roads of mobile technology.

In our interview, Williams described Antoine Wright to me as "the granddaddy of mobile ministry," a moniker that seems silly when you see Antoine—a young-looking thirty-eight-year-old—but appropriate given the speed in which technological change has happened in the past twenty years. People in the mobile ministry scene have dubbed Wright "the granddaddy" because he was the first to start publicly advocating for bringing mobile, digital devices into missionary work through his online outlet *Mobile Ministry Magazine*, which ran from 2004 through 2016. When I spoke with Wright in 2017, he was wearing AirPods that connected to his iPhone, an Apple Watch, and a ring that doubled as a GPS-enabled fitness tracker. He told me he had his Spectacles—Snapchat's video-capturing sunglasses—inside. Wright has always been an early adopter of technology; he said he owned a PalmPilot when it was commonly mistaken for a Game Boy. Around the year 2000, in a church, Wright was looking at his PalmPilot and thought to himself:

What's going to happen when everything that this leader, this pastor, or teacher is saying in front of me is accessible in this handheld device? Will that person matter? How does that change the threading of this community that I've come to love, appreciate, or know and: Oh crap! Are we ready for that as a local community? Are we ready for that as a faith community, and oh my God, is Christianity, Judaism, Buddhism—is anybody ready for that? (Wright, 2017)

These initial questions led Wright to start writing the magazine and to assemble a network of church and tech professionals interested in how to use new media technologies for Christian evangelizing and community building. And with that, Wright became the mouthpiece of a growing movement of evangelicals interesting in using mobile devices, as they put it, "missionally."

For most evangelicals, using technology "missionally" means using digital tools to spread the gospel or to connect individuals into digital networks that ideally become virtual communities. Wright, however, has a more expansive view of mobile ministry. As he sees it, the world is becoming more and more connected, which requires a more mobile definition of faith itself. "A decade ago," he explained to me, "it was disconnected first connected later, but if the default is connected and if the default means that we are connected, we are probably more like plants and less like rocks—every plant in the world is connected at the root system" (Wright, 2017). In his view faith has to be a lens through which believers can interpret an increasingly connected culture. And because of this, he sees a potential Christian future looking more like the distant past, as in the past of Medieval Europe, when "the thing that made Christianity valuable is that Christianity was able to spread because you could embed it into every area of someone's life. It was in art if you were a noble you literally hummed hymns or creeds while you were working in the field it was very much a part of who you are. I believe we get back into that" (Wright, 2017).

Wright believes faith has to infuse every aspect of life in order to succeed in a connected world; it has to be part of every connection, it has to create connections, it has to be immersive and ever-present in believers' lives. In other words, it has to be on your iPhone. For Wright the challenge that the church faces is understanding how to embed Christian directives in ever-present mobile phones and other consumer media technologies, and he views this imperative within a global framework. Not only does Wright fear losing a new generation of American millennials to agnosticism, as others I spoke to for this book also fear, but he fears that global Christianity could lose out to other religious voices. "If the Christian church drops the ball," Wright (2017) predicts, "and generally speaking another religion comes in and picks up the ball—that's just kind of the way history has happened—and it's like crap, now we are marginalized, now we are back to being the little church in the corner trying to figure out what our beliefs are."

Wright hopes that Christians can avoid this future by taking to technology and understanding the world as a vast, digitally connected network of believers. His vision is not too distant from the vision of the man he calls his "intellectual mentor," Howard Rheingold. In *Virtual Community: Homesteading on the Electronic Frontier* (1993), Rheingold wrote glowingly of the potential for communities to spring up on the internet, using his experience with the online community WELL as an example. And like Wright, Rheingold sees the importance of having knowledgeable stewards of this technology deeply understand it in order to use the technology to enhance democratic discourse. For Rheingold (1993), in the context of democracy, a more connected world could be utopic or dystopic because:

> technology offers a new capability of "many to many" communication, but the way such a capability will or will not be used in the future might depend on the way we, the first people who are using it, succeed or fail in applying it to our lives. Those of us who are brought into contact with each other by means of CMC technology find ourselves challenged by this many-to-many capability—challenged to consider whether it is possible for us to build some kind of community together. (p. 12)

Wright takes Rheingold's simultaneously optimistic and precautionary stance when describing the connected world, although he applies this lens to the future of Christianity rather than the future of democracy. Wright hopes that the ubiquitous connection that the New Roman Roads facilitate will change Christianity for the better, but he also fears that if Christians are unwilling to change and adapt, it could change them for the worse. In either vision, change is inevitable, the technology cannot be ignored or willed away; it has to be considered, and ideally, "stewarded" correctly.

THE OPEN CHURCH

Although they are optimistic about digital technologies, unlike evangelicals working in the faith-tech sector, Christians promoting technological tools in missionary organizations, many of which are nonprofits, do not as fervently believe in the power of tech businesses. My interviewees often pointedly critiqued what they saw as the increasing corporatization of the internet. For example, John Edmiston (2017) prides himself on having been one of the first people to take to the internet, even before the introduction of the World Wide Web. He tells me that he had the first prayer website on the internet, the first Bible teaching website, and the first Christian dating website. Yet, despite his enthusiasm for the early web, Edmiston is now cautious, because "it has been massively corporatized so that Facebook and Google and a few companies basically own the internet" (2017). He went on to explain his concerns about corporate surveillance, about the impossibility of being forgotten, and about problems with unmasking of missionaries on social media. For Edmiston, who had seen the early potential of the internet in the 1980s and 1990s, the rise of Facebook was an example of problematic corporatization happening in the tech space. Others echoed these concerns. Although Antoine Wright (2017) was typically optimistic about the trajectory of technology, he told me: "I'm also as scared as everybody else because I read the terms of service for everything, so I understand the nature of surveillance culture."

While the culture of faith-tech emulates start-up culture and sees stylish entrepreneurs extolling the potential of redemptive entrepreneurship, the mission field is more likely to attract self-proclaimed "geeks," "techies," and "hackers." Because of this, evangelicals in the mission field who emphasize technology often quoted people like Richard Stallman and Howard Rheingold rather than Sergey Brin or Steve Jobs. They prefer the early visionaries of the internet who believed that the connective power of the medium had utopic potential. My interviewees especially expressed affinity with the Free and Open Source Software movement (FOSS). FOSS refers to a long-established computing community that creates and contributes to software with an open source code. This type of software allows hackers and computer enthusiasts to freely copy and improve upon digital products. The ethic of sharing that is the basis of the FOSS movement was clearly articulated in the 1980s by Richard Stallman, who laid forth a set of moral rules for computing. For example, "As a computer user today, you may find yourself using a proprietary program. If your friend asks to make a copy, it would be wrong to refuse. Cooperation is more important than copyright" (Stallman, 2010, p. 52).

In her ethnography of hacker culture, Gabriella Coleman (2013) explained that FOSS gave a generation of young men (and to a much lesser extent, women) who were fascinated by computing a rallying point. The software provided a chance for these proto-hackers to tinker, and in so doing, to take ownership of the process of computing. It also allowed for a community to develop that took FOSS as its lingua franca and that began to take the ideology of sharing as its central moral tenet (Coleman, 2013, p. 26). Both within and outside of the hacking community, FOSS has been taken as a touchstone for a variety of ideological positions. Stallman conceived of it in a socialist framework, but others have read it as a libertarian ideal (see Raymond, 2001) or as a Marxist ideal (see Wark, 2004). The Christian missionaries who express enthusiasm for FOSS see it in biblical terms. Yet uniting these fractions is an imaginary that challenges the dominant neoliberalist capitalism that predominates in the tech world today. The collective production that FOSS programmers perform challenges received cultural understanding of the necessity of Fordist and post-Fordist production models.[7] Because it is a challenge to dominant market modes, the logic of FOSS extends beyond just software. It allows for "a form of collective action that has crystallised in response to capital's quest to commodify ideas, knowledge and information" (Berry, 2008, p. 101). FOSS is thus diametrically opposed to much of the tech sector, which relies on a proprietary understanding of intellectual property to turn a profit.

Many of the people I spoke with for this chapter had worked on open source platforms, and they echoed the rhetoric of the FOSS movement precisely because it challenged dominant economic models. For example, Antoine Wright (2017) said: "The open source community and the Bible sharing community have seemed to almost work lockstep at least in the last decade they did in terms of making content available for least of these." For Wright it is the fact that the FOSS community produces content geared toward older devices that proves their idealistic goals, the same goals shared by missions organizations who hope to proselytize to "the least of these," the oft-used biblical term for poor and marginalized people.

Life.Church has also incorporated the ethics of the FOSS movement into their mission. On their expansive website they have a section called "Open" that offers church resources for free download. Life.Church sees its work as instrumental to an imagined global church community united under the authority of God. The lead pastor of Life.Church, Craig Groeschel, has written about the impetus behind "Open" recalling, "What if, we asked idealistically, we just gave away our creative content—for free?" He describes this as a breakthrough in Life.Church's approach to ministry, and he immediately conceived of it as "open sourcing"—a concept he traces to "a few computer companies who started sharing their in-house software with anyone who wanted it, trusting (hoping) that others would follow" (Groeschel, 2007). Groeschel sees this as an ethic common in the tech world, but he also places it in a distinctly Christian framework, explaining that "open sourcing is giving away the rights to use what already exists. In the ministry world, we could define it as giving away what wasn't ours in the first place. (It all came from God, anyway)" (Groeschel, 2007). Here, he takes the concept of "open sourcing" and places it in a theological context in which God is the ultimate rights holder, not an individual or a church. Life.Church's online software, which they offer for free on Open, has become so popular that it likely has put other for-profit faith-tech platforms out of business.[8]

Johnathan Pulos runs a nonprofit called Missional Digerati, another organization involved in the Mobile Ministry Forum. He told me that the idea for Missional Digerati was sparked by an eighteen-week ministry course called "Perspectives on the World Christian Movement." At the time Pulos took this course, he was working as a programmer in a large company. The course prompted him to imagine that his training as a programmer could be used to create an organization that was basically a full-time hackathon, where programmers and engineers could get together to create technology to support missions work. "I've always dreamed about the idea of Missional Digerati becoming a think tank where you get these techies together [and] their heart and their mission is fulfilling the Great Commission, and they've got all the resources they need on hand to just do stuff: build the right projects, test it out, see the results" (Pulos, 2017). He explains that this was not a business idea—the goal was not to charge anything for these services.

In this way Pulos exemplifies the ethics of the mobile ministry community, where making money is beside the point. The organizations that I looked at were all nonprofits, but beyond that status, the people involved, like Pulos, seemed to regard money and money-making as merely a hindrance to their ultimate goals; they see themselves as missionaries and welcome the self-sacrifice that role requires. In the five years since Pulos has been running Missional Digerati, that dream has not totally been realized; he now mostly works with missions organizations, consulting with them and building their technology. But it remains important for Pulos to encourage his clients to open source everything that he produces. For him, open source software is the key to his goal of having a frictionless innovative atmosphere of Christian technologists working to evangelize the globe.

Tim Jore started his career working in Bible translation, and in tandem he participated in the open source movement. Because of this, he recalls, "I learned how the community works, how very different it is from commercial products, and really started to internalize some of those principles and more importantly started to recognize that much of it is, I believe, deeply rooted in biblical principles" (Jore, 2017). Like Pulos and Life. Church, Jore spins the ideology of FOSS in a Christian way, seeing the principles of sharing as biblical ones. He also read Lawrence Lessig, who applied the logic of FOSS to cultural production. In 2001, Lessig created Creative Commons, a nonprofit organization whose mission is to create and maintain open licenses specifically tailored to the concerns of the digital age. Creative Commons translated the ethics of the FOSS movement to cultural production, providing the impetus for what is known as the Free Culture movement.[9]

The ideas Jore gleaned from FOSS and the Free Culture movement eventually prompted him to write *The Christian Commons* (2013). In it, Jore describes being a missionary in the field in Papua New Guinea and trying to use software to help local Christians translate the Bible into their dialect. "Everything had been going fine until the software installer prompted us for a license key to translate the Word of God," he recalls, "but since we did not have a license key, we could not proceed with the installation of software" (Jore, 2013, p. 15). This experience was significant for Jore (2013) because "in the legal context of 'all rights reserved,' the global church is unable to work together without restriction or hindrance to leverage Internet and mobile technology to the fullest for the purposes of God's Kingdom and the equipping of His Church" (p. 14). Jore's book describes an alternate paradigm in which Christian technologists, translators, and publishers might work together to create a free and open Christian commons that might be accessed by anyone.

When I caught up with Jore in the fall of 2017, he excitedly related that "what had been described mostly from a theoretical standpoint in *Christian Commons* in 2012 is just accelerating globally and it's thrilling to be a part of" (Jore, 2017). He works in innovation and strategy at an organization then called Distant Shores (later the name was changed to UnfoldingWord), which has developed a model that allows indigenous Christian communities to get access to biblical resources of all types in "gateway languages" that they can then use to translate into local dialects and spread throughout their region. All of the media Jore transmits is open through Creative Commons licensing, meaning that it can be shared, remixed, and rebranded in any way the user sees fit.

The most pushback Jore has received on his ideas comes from publishers and established Christian organizations. For example, he expresses frustration that the Greek New Testament has been protected by publishers who have copyrighted their translations, in effect making them inaccessible for people who cannot pay for them, such as Christian pastors in the developing world. Although evangelical outreach strategies have historically been based on giving away free things, especially Bibles, Christian media producers and distributors have also been involved in a balancing act between giving away cultural products and retaining ownership of the right to these products. Because of this, there have been controversies about how churches should interpret copyright law in regard to sermons (see Smietana, 2014), Contemporary Christian Music (see Gormly, 2003), and other cultural forms (see Berg, 2003). The Christian intellectual property regime established in the 1970s, 1980s, and 1990s has proven to be profitable for Christian businesses; Bibles especially are a source of millions of dollars in revenue for a variety of Christian organizations every year.[10] But Jore, inspired by what he believes are the biblical principles of the FOSS and Free Culture movements, believes that to truly steward technology, Christians have to get away from a paradigm that has valued business and money-making over sharing and unfettered circulation.

For Jore, the Open Church movement is also a paradigm shift because it gives more power to those in Indigenous populations in the Global South to control content. As he explains:

> Perhaps the most important aspect in all of this is getting away from the idea that we are going to come up with the best solution here and then we are going to ship it globally. That's more of a commercial model and that's fine. It's a business model, that's great. And sometimes it's extremely effective, but where I'm going with this is very similar to our take on the content and the licensing. We are very interested in letting the church have the freedom to solve their own problems. Not in a disconnected way, not in a this is your problem you take care of it, but in the sense of let's give everything that we've got with the freedom so that everyone can be creative to meet the needs together. (Jore, 2017)

Jore explicitly sets his work apart from commercial models of technology and media production. Instead, he sees technology as a tool that can be used to empower churches around the globe, and he understands that means relinquishing some control.

Since I spoke to Jore in 2017, there have been shifts in how large evangelical organizations and publishers have understood the intellectual property norms that hindered open sharing of Bibles and other theological resources. Biblica, for example, has released an "Open Bible" that they have licensed using an Attribution-ShareAlike Creative Commons license. And the Gospel Coalition has also begun to release theological resources under the same Creative Commons license. The influence that those in favor of more generous licensing have had with publishers and organizations seems to have had an effect on how evangelical organizations conceive of intellectual property. Through open technology, open resources, and a connected globe, Jore and others imagine new nodal points in global Christianity emerging. Jore believes he has seen the beginning of this in parts of India, where sharing, remixing, and circulation of Christian content has accelerated in the past few years. He compares this to a flame hitting gasoline: it cannot be stopped.

A GLOBAL PARADIGM SHIFT?

On the surface Jore's understanding of how to translate media messages through Indigenous cultures by giving those cultures control and opening content to remixing and sharing without restriction is reminiscent of the anthropologist Faye Ginsburg's (2008) view of Indigenous media production in the "digital age." Ginsburg urged scholars to think beyond the common argument that points to the problematic existence of a vast digital divide, and instead introduces case studies that show that Indigenous media producers often have different views on how technology might be used than those prescribed by Western observers. She shows that Indigenous producers have especially critiqued the hegemonic insistence on Western intellectual property norms. For Ginsburg this is important to keep in mind when we refer to the "digital age" as a unified ideological moment because, as she writes, "the language smuggles in a set of assumptions that paper over cultural difference in the way things digital may be taken up—if at all—in radically different contexts and thus serve to further insulate thinking against recognition of alterity that different kinds of media worlds present" (Ginsburg, 2008, p. 129).[11] Digital habitus is culturally constructed and is always influenced by the historical realities of places—some cultures take to the internet and see it as a central outgrowth of their natural social communication, while others do not.[12]

Following a similar line of thinking, some organizations in the Mobile Ministry Movement are dedicated to using digital tools explicitly to empower Indigenous communities to create their own products and content geared toward spreading Christian messages. Code for the Kingdom, for example, hosts weekend-long "hackathons" all over the world because they hope to be "igniting the Christian passion and purpose of technologists and entrepreneurs to innovate culture shaping technologies that would reclaim our times for the gospel" (Code for the Kingdom, 2017). They have hosted events in places such as Addis Ababa, Bogota, and Jakarta as a means to engage Indigenous media producers. Another organization called Indigitous is also illustrative. A portmanteau of "Indigenous" and "digital," the organization explains that both of these concepts are crucial to their mission. "Indigenous because it reflects our desire for locally-generated strategies that work in each generation, language and culture. Digital because we believe digital tools, resources, platforms and strategies can accelerate God's mission" (Indigitous, n.d.). To this end, like Code for the Kingdom, Indigitous sponsors hackathons around the world, hosts a global network of volunteers, and offers social media training to Christians globally.

I spoke with two Indigitous volunteers, Jeyanti Yorke and Janakan Arulkumarasan, a married couple who live in Singapore. They have also created their own app called Oikos, which offers church management services to churches around the world on a sliding pay scale. Arulkumarasan told me that the Western, and especially American, focus of Christian technology production is a problem because it excludes non-Westerners from the means of production. Arulkumarasan believes that this deficit in the balance of technological production exists because of the structure of the global economy. He stresses that "there's no lack of innovation in the rest of the world, in the majority world. We have been to places in Ethiopia where they have internet for half an hour a day and they are able to code things at the same level as these American developers. The difference is they don't necessarily have access to market" (Arulkumarasan, 2017). Because they do not have access to market, and thus cannot easily sell their products, they do not have a voice in how technology is produced. Creative Indigenous producers are in effect silenced.

In one sense, the way that Jore, Code for the Kingdom, Indigitous, and others in this movement have reimagined how media might be used in Indigenous contexts follows from Ginsburg's understanding of alternate media worlds. Because Indigenous producers can openly use content in any way that they see fit, they can, for example, put their own branding on a teaching video, or they can add their own biblical exegesis to a openly shared cultural product, these Christians believe they are ceding their power as theological and cultural gatekeepers to Indigenous communities. Yet, as Arulkumarasan alludes above, the problem goes beyond access to materials and tools. This is a problem that comes into many programs constructed in a Western context that hope to aid the majority world, and a focus on technological production and technology use may enhance that problem. In his book on the development of the iPhone and the app economy, Brian Merchant (2017) visited Nairobi's "Silicon Savannah" and found that although Western technology companies and NGOs had enthusiastically pushed local producers to create technological products as a means to bring money into the region, these Western organizations had not understood how entrepreneurship worked in Africa.

"The idea of mobile revolution or an app-based revolution ported poorly from the U.S. or Europe," Merchant wrote, "where it was a cultural phenomenon, to Kenya, where the reality was much different" (2017, p. 179). And Yuri Takhteyev's (2012) study of coders in Rio reveals the complex relationship that those working at the periphery have with those in the center, in Silicon Valley. Takhteyev's work also points out that technology production and especially open source software presents a high barrier of entry for those programmers that do not speak English, as the dominant coding languages all rely on English and FOSS particularly uses English as its lingua franca. "Participation in open source projects involves a complex negotiation of culture, language, and geography," he wrote, "and is often *harder* than engaging in other forms of software practice, since it requires *more* fluency in foreign culture and demands *more* of the resources that may be hard to find" (Takhteyev, 2012, p. 9, emphasis in the original). Thus, porting the idea of a "hackathon" to Bogotá, may not be as empowering as the people who run these organizations hope it to be. Similarly, cultural products initially produced in the West may have a high barrier of entry for Indigenous people, or may not be relevant in Indigenous cultures.

Although mobile technology has been heralded as the means by which the world might truly be united in a McLuhian "global village," the way that new technologies have been received and produced in local places belies this idea. Ultimately, then, the movement of digital evangelism is still one-way—from the West to the rest. Now not only are the messages designed in the West, but so are the technologies used to transmit them. Like the iPhone, they are "designed in California, assembled in China." As

Bill Wasik (2015) has pointed out, "the smartphone—for all its indispens-
ability as a tool of business and practicality—is also a bearer of values; it
is not a culturally neutral device." Mobile technology allows Christians to
imagine themselves as part of a globally connected network of believers
and allows for predictions of new global nodes emerging to challenge the
historical Western dominance of Christianity. But because these technological products are designed in a Western, capitalist context, they have in
effect excluded the majority of the world from their production.

Langdon Winner (1986) famously asked if artifacts have politics, and
the mobile phones that have found their way into the hands of people
all around the globe certainly do. The politics that these devices carry
with them are not neutral, and although evangelicals have tried to tinker
with intellectual property norms, with software, and although they have
hacked and innovated, they are still dealing with and sometimes export-
ing Western cultural products. And in this way these Christians are falling
into the same trap as their counterparts in the technology industry who
see technological spread and the closing of a global digital divide as the
central problems of the "digital era." To return to Faye Ginsburg's critique,
in the spread of these culturally, historically situated technologies is the
spread of the ideologies that accompany them, and while many Christians
are trying to reimagine and rethink these hegemonic views of technology,
because they rely on these tools, they face an uphill battle.

CONCLUSION

In reimagining the globe and they ways they might redeem it, the evangelicals involved in the mobile ministry world are also reimagining themselves and their role in a vast network connected by the "new Roman
roads." Unlike those in the faith-tech space, however, these Christians are
not out to make money, but rather want to spread the gospel and they see
technology as a prime way to do that. At the same time, their technological influences are those like Richard Stallman, who want to keep technological products free and open, and as such these evangelicals explicitly
and implicitly critique the neoliberal economic principles that govern
the commercial technology industry. While some hope that a connected

planet will shift power dynamics and unite the global church, the rhetoric espoused by many in the mobile ministry movement continues the problematic rhetoric that has infused missionary work for hundreds of years. It also reveals the problems of a technologically deterministic discourse inherited from techno-utopianists.

Perhaps more important, the Christian missions field has historical baggage that it carries with it into the digital age. Although the missionaries that I spoke with tend to see themselves as John Edmiston does, as " a servant to the majority world" (Edmiston, 2017), and although many spoke about the problematic assumptions that have been built into missionary work, there is still ample evidence that the norms of mobile ministry are Western, especially American. Furthermore, by emphasizing the importance of technology and borrowing the ideologies of techno-utopianists like Howard Rheingold and others, evangelicals may in fact be intensifying the Orientalist leanings of Christian missionary work that Christian ecumenicism has been trying to fight against since 1910.

Digital habitus has changed American evangelical culture, and it has changed the way that many American Christians imagine global outreach. However, this habitus is culturally situated and may not translate into other spaces as readily as many evangelicals hope. Yet those missions-focused evangelicals who believe that digital habitus could bring about a global power shift notice something inherent in the technologies that define the so-called digital age. Digital habitus is a threat to established power structures because it allows people to bypass cultural gatekeepers. Ordinary Christians now have access to powerful broadcast technologies that allow them to create their own audiences and publics. The next two chapters chart how evangelical power structures have been challenged by new celebrities who are using social media and podcasting to reveal and reckon with what they see as the toxic norms of American evangelical culture.

4 The Influencers

THE RISE OF EVANGELICAL INFLUENCERS
AND THE POTENCY OF POPULAR PAROCHIAL
FEMINISM ON SOCIAL MEDIA

I was shaking as I stepped into my front yard and told my husband the news: Rachel Held Evans had died. I had been following updates on her condition since she was hospitalized for what seemed like a routine illness. Things quickly took a turn after Evans seemed to have an allergic reaction to antibiotics and was placed in a medically induced coma. Like many of her longtime fans and followers, I had been refreshing her website multiple times per day since I heard she was ill.

Evans's death came as a shock to me and to many others. She was only thirty-seven, had two small children, and had just tweeted about *Game of Thrones*. Messages of remembrance poured out onto the internet from such cultural luminaries as the director Ava Duvernay, the folk band The Mountain Goats, and Russell Moore, the former head of the Southern Baptist Convention (SBC). Tributes and obituaries in the *New York Times* and *The New Yorker* followed, and Evans's book *Searching for Sunday* made it on the *New York Times* bestseller list for the first time, nearly four years after its 2015 release. But perhaps more interesting than the high-profile praise she received after her death was the response from ordinary women on social media. Using the hashtag #BecauseofRHE, women expressed the ways that Evans had changed their life and faith. One wrote:

"Today I'm missing the church she built and her pastoral care. She was my pastor. Her Twitter mentions and blog comments were my congregation."[1] Another wrote: "I found my voice bc she paved the way." And one more claimed: "An entire generation of theologians and ethicists had risen up, blooming from the seeds of the words she leaves behind."

Evans was baptized and educated in evangelical institutions, and she wrote about the social pressures of being a young evangelical woman whose questions about theodicy, eternal damnation, and the culture wars were quashed by her church community in Tennessee. When her church became involved in an anti-LGBT political campaign, Evans and her husband decided to leave their community and begin their exit from evangelicalism. Although Evans did not take this decision lightly, she explained: "It is perhaps no coincidence that I discovered blogging around the same time, and along with it, a whole community of people from across the world who smiled back at me from the tiny avatars in the comment section and bestowed upon me, like gifts wrapped in delicate paper, two very powerful words: *me too*" (Evans, 2015, p. 62). Although she had left the community of her youth, Evans found a new community that shared her frustrations, hopes, and fears through blogging and later on social media. And from this perch, Evans became a powerful native critic of evangelical culture.

Evans was one of the many Christian women who successfully navigated the blog circuit of the early 2000s and parlayed her popularity onto social media. Because she was a woman with strong opinions about evangelical culture and theology, she was both lauded and maligned depending on where you sat in Christian culture. Using social media, Evans was able to marshal a public comprised of women who identified and related to her story and responded with their own. "Me too" was what Evans heard from her followers, and later this was the phrase that would erupt on social media, burying in its wake once powerful abusers, and once unmovable cultural norms. Digital habitus has made natural what was once considered strange: vast communities of strangers who never meet each other and yet share deep ties. These connections have sparked social movements all over the world, and over a short period they have reshaped the evangelical discourse about sex and gender.

This chapter argues that evangelical women have used the affordances of social media to promote themselves and in so doing have created a new form of Christian feminism that I call popular parochial feminism. As they have gained followers, these influencers have mobilized a counterpublic of Christian women who have used the platforms at their disposal to start new conversations about misogyny, abuse, and sexism in evangelical culture that have gone beyond the internet and have rocked the American evangelical power structure.

NEW CHARISMATIC AUTHORITIES: THE CHRISTIAN INFLUENCER

In a professionally produced eight-minute video posted on YouTube that has been viewed over forty-two million times, a beautiful, young, blonde bride reads her vows to her groom: "You have so many dreams and passions about changing and helping the world in Jesus's name. I can't wait to wake up every Sunday from this day forward and go to church and worship our King with my best friend" (The Labrant Fam, 2017). As the video continues we see close-ups of impeccably designed, abundant flower arrangements, drone shots of a beautiful orchard venue, and handheld shots of perfectly styled children running joyfully in tailored dresses and suits. This video—a Pinterest wedding board come to life—documents the wedding of Cole and Savannah LaBrant (Cole and Sav), YouTube stars in their own right who merged their families and brands when they married in 2017.

In addition to Savannah's first child from a previous relationship, Cole and Sav have two children together, Posie and Zealand. Their YouTube channel is populated by wholesome vlogs in the style of a reality show that begins with shots of the "stars" of the show. Cole, Savannah, and each of their children have their own Instagram accounts, each of which boasts more than a million followers. Savannah has 6.4 million followers on Instagram and their family YouTube channel has more than 3.5 billion views. In their book *Cole + Sav: Our Surprising Love Story* (2018), Savannah describes her career as an influencer as accidental: "In the Instagram world, if you upload nice, clean pictures and if you have a good amount of followers, stores and others will send you outfits to wear in your pictures. It's basically free advertising for them" (p. 10). The way she tells it, she was just a pretty southern California blonde with a cute, photogenic daughter, which translated into free products. When Cole came along—a young, blonde YouTuber from Alabama—their romance and marriage garnered them even more followers. The LaBrants have successfully monetized their vision of Christian domesticity; they now live in a multi-million-dollar mansion in Orange County, California, complete with tennis and basketball courts and a pool. For many of their followers the LaBrants represent #RelationshipGoals. They are wholesome, attractive, wealthy, young Christians who post vlogs in which they goofily play with their kids in their pool and choreograph dance numbers with multiple outfit changes.

As an evangelical influencer, Savannah LaBrant represents the normative vision of Christian femininity that has become popular on social media. Although she wants everyone to know that she has not had a perfect life—she speaks tearfully of her previous failed relationships and her parents' divorce, for example—Savannah fits perfectly into the frame of evangelical womanhood as well as the what the journalist Taylor Lorenz (forthcoming) has called the new American dream of Generation Z—influencer status. In LaBrant's case her popularity has translated into lucrative deals with brands that want her audience to associate their products with her wholesome lifestyle, and she now promotes her own clothing line. But in the case of other female evangelical influencers, their popularity has made them leaders in evangelical culture and has translated into conference tours and best-selling books.

In 1956, Donald Horton and Richard Wohl gave the strangely intimate relationship we have with media personalities a name: parasocial interaction. They were studying the then-new medium of television—a medium that seemed to invite interesting, colorful characters into the living rooms of ordinary people at scheduled times throughout the day. Though we did not really know the people populating our screens, we could rely on them to be there, we could fantasize about them, we could read tabloid articles about their real lives, we *felt* that we knew them. The one-sided relationship we developed with these characters was not quite social, yet it was still real. In the age of digital habitus, parasocial relationships have edged further into the social world, and celebrities have started interacting with their audience, or followers. Celebrities have invited their followers into the intimate spaces of their homes, and they post #nomakeupselfies, pictures of their children, and pictures of their food. The more *authentic* we believe them to be, the more we think we are seeing their real, unfiltered selves, the more followers these celebrities gain.

The social media economy runs on followers and likes and encourages users to develop a brand—a consistent frame of themselves and their lifestyle that they can maintain to help them cultivate an audience. Influencers and influencer wannabes do this in a variety of ways—for example, they use hashtags to index the populations they want to reach. They use professional photo equipment to perfect their images. They emulate both celebrity culture and marketing techniques gleaned from corporate brands.[2] And this all takes much more work than it seems because to be successful, self-branding must be somewhat invisible.[3] That is, if it is clear that a social media celebrity is trying to promote themselves in their social media posts, they risk losing that crucial air of authenticity—and authenticity is akin to currency on social media.[4] Savannah LaBrant's success, then, was not accidental, as she claims in her memoir. Rather, it was the result of a careful and calculated understanding of how to leverage her self-brand to gain followers and sponsors.

But even when monetizing their self-brand is an end in itself, social media influencers still post content that is distributed to their followers, and this content is read and seen by large, distributed networks of people. Because of this, successful Christian influencers also have a type of authority—their words and actions hold weight to their followers who see them as people they might aspire to be. In his work on the subject Max Weber (1968) divided authority into three types: rational or legal authority, traditional authority, and charismatic authority. Weber wrote that charismatic authority relies on "devotion to the exceptional sanctity, heroism or exemplary character of an individual person, and of the normative patterns or order revealed or ordained by him" (1968, p. 215). Popular Christian influencers like Savannah LaBrant have created self-brands on social media that employ an understanding of what it means to be an "authentic" Christian woman, which in itself is a normative frame. What is read as "authentic" also serves to display and police the boundaries of modern Christian womanhood. And this is why women who become influencers are so carefully watched, and sometimes harshly reprimanded in evangelical culture. They have a particular cultural power over their followers, who see them as authentic, who see them as #goals, who see them as authorities. And in evangelical culture, women's voices have historically been carefully controlled by formal and informal authoritative structures that typically do not allow for women to preach at the pulpit.

Many evangelicals consider what is known as "egalitarianism" to be theologically and doctrinally radical. "Egalitarians" are those Christians who believe that women and men are equal, and although they may be gifted with different abilities, they should have equal access to power and authority within churches and families and should be encouraged to find their own paths in the workplace and in society. "Complementarians," by contrast, believe women and men are meant to fulfill distinct roles in culture and in families; that men are meant to lead and women to support.[5] Complementarianism—also referred to in more colloquial contexts as "male headship"—tends to be the accepted position in most traditional American evangelical churches[6] and in evangelical culture more broadly conceived. Rachel Held Evans described reading Christian marriage books that taught that "if the right person leads (the man) and the right person follows (the woman), if one person makes the money (the man) and the other person keeps the home (the woman), if there is one protector (the man) and one nurturer (the woman)—then everything will work out" (2015, p. 241). In Evans's experience the gender roles in evangelical culture were explicit and manifested in a panoply of social pressures.

Peter Glick and Susan Fiske (1996) have theorized what they called "benevolent sexism," an attitude borne from the idea that the feminine is a sacred category that must be protected and controlled by men. This sexism works to control women by casting them in the role of the subordinate and by punishing—either socially, psychologically, or physically—those women who trespass these boundaries. Complementarianism is one permutation of benevolent sexism. The complementation gender structure is sometimes diagrammed using three umbrellas. The man falls under the umbrella of God's authority and his umbrella of authority covers his wife, whose authority covers the children and home. Many evangelicals believe these roles to be prescribed by the Bible and thus women who step out of their roles are sinful, or even heretical. When Christian women become influencers, they exercise a power outside of the home and rearrange the umbrellas. In a Fathers' Day Instagram caption directed at her husband Cole, Savannah LaBrant expressed what seems to be a traditional Christian understanding of complementation marriage: "Thank you for leading us, loving us, protecting us, and taking care of us every single day" (LaBrant, 2018).

Yet as Savannah LaBrant uses her role as a traditional wife and mother to gain followers, she enhances her own brand as well as her charismatic authority. Christian influencers can and do use their charismatic authority to "preach" on Instagram, on Twitter, on their own blogs, and through their best-selling books and sold-out conference tours. But they also have to walk a tightrope to maintain their authenticity and their Christian bona fides.

THE POPULAR PAROCHIAL FEMINISM OF THE CHRISTIAN INFLUENCER

Rachel Hollis's *Girl, Wash Your Face* (2018) is a high-energy self-help book geared toward women. The book was published by the Christian imprint Thomas Nelson and went on to become a *New York Times* best seller and one of Amazon's top selling books of 2018. The cover of *Girl, Wash Your Face* is a photo of Hollis, smiling, wearing jeans and red Converse sneakers, getting sprayed by a yellow fire hydrant. It is meant to display the book's gist: Hollis is not perfect, but she's happy and fulfilled despite her sometimes complicated life. Each short chapter of this self-help book is titled with a "lie" that women tell themselves, and Hollis ends each one with three takeaways to help women overcome this lie in the style of the self-help genre. In the prologue Hollis talks about the lies her book is preoccupied with and explains that they are "perpetuated by society, the media, our family of origin, or frankly—and this in my Pentecostal showing—by the Devil himself" (2018, p. xii).

Hollis's writing is honest and funny, and she talks frankly about drinking too much and occasionally popping a Xanax. She tells her readers that she peed herself while jumping on a trampoline and that she shaves her toes. She writes about the importance of the female orgasm. But she also reveals some of the dark periods of her past. She shares about finding her older brother's body after his suicide and, later, about the destructive coping mechanisms she used to work through her pain. Hollis describes these central traumas of her life as though she is talking to an old friend. These revelations endear Hollis to her audience who see them as markers of authenticity. Since the success of *Girl, Wash Your Face* made Hollis an international celebrity, she has written *Girl, Stop Apologizing* (2019) and *Didn't See That Coming* (2020), which chronicled her divorce from Dave Hollis, who has parlayed his ex-wife's success into his own media career and become famous in his own right.

Rachel Hollis first became Instagram-famous when she posted a photo of herself in a bikini that showed off her stretch marks on Instagram. In the photo she is standing on a beach, smiling widely and tousling her hair. Though she is fit and tan, there is a bit of a pouch around her belly button. For many of her female followers, this photo was proof that Hollis was showing her true, un-retouched self. It is a photo that women, especially those who have had children could relate to and empathize with, and it shows her followers that Hollis is authentic and open. In the caption that accompanies the picture, she explains:

> I have stretch marks and I wear a bikini. I have a belly that's permanently flabby from carrying three giant babies and I wear a bikini. My belly button is saggy...(which is something I didn't even know was possible before!!) and I wear a bikini. I wear a bikini because I'm proud of this body and every mark on it. Those marks prove that I was blessed enough to carry my babies and that flabby tummy means I worked hard to lose what weight I could. I wear a bikini because the only man who's [*sic*] opinion matters knows what I went through to look this way. That same man says he's never seen anything sexier than my body, marks and all. They aren't scars ladies, they're stripes and you've earned them. Flaunt that body with pride! #RealAndChic #HollisHoliday (Hollis, 2015)

With this photo and the accompanying caption, Hollis invited spectators to view her imperfect body and put on display the markers of her Christian womanhood: in this case, the bodily markers of past pregnancies. The caption emphasized the importance of being a mother and a wife to "the only man who's opinion matters"—her then husband. But this photo also displays markers of self-care and markers of wealth—a bikini embroidered with Hollis's initials, an expensive beach vacation—and these too provide a semiotic index of aspirational connotations. Though her post caught attention because it showed a woman with stretch marks, it also presented a particular class habitus that is central to how Hollis and other influencers brand themselves. As in Savannah LaBrant's perfect wedding video, the world of the influencer has to look somewhat better than the real world. Followers must aspire to live like or be like influencers. That is what makes the influencer influential.

The perception that Hollis's is a particularly successful vision of modern Christian womanhood has granted her charismatic authority. She headlined a suite of conferences she branded "Rise" conferences targeted at empowering women in a variety of ways: in business, in their marriages. On these national tours Hollis routinely sells tens of thousands of tickets. In her writing, on her social media platforms, and through her conferences, Hollis encourages her mostly female audience to empower themselves: to succeed in their careers and become leaders. In her book Hollis (2018) tells her readers that "working women sometimes have to fight their way through patriarchal systems" (p. 130), and she lists the social pressures that women face in their communities that might make them more likely to pursue traditional motherhood rather than careers. "It makes me wonder," she writes, "how many women are walking around living in half their personality and in doing so, denying who their Creator made them to be" (Hollis, 2018, p. 130). Here Hollis names and shames patriarchal systems of oppression and claims that women who follow their dreams and become successful are in fact following God's plan, even if their ambitions put them at odds with the social pressures they face in their communities. This goes against the grain of traditional complementarian thinking, which asserts that the woman's role should fall squarely in the realm of the home.

I argue that Hollis and other Christian influencers like her are feminists, but because they speak in the register of evangelical culture, the broader world might not recognize them as such. I term their feminism "popular parochial feminism." The popular parochial feminism of the Christian influencer exists alongside and within the frame of benevolent sexism as it is a feminism that values the role of women as caregivers, wives, and mothers. Although it allows for women to express their voices, their power, and perhaps most important their role as economic agents or money-makers, this feminism still holds traditional understandings of femininity and especially motherhood as sacred categories. Women are not free to pursue any lifestyle that suits them. For example, they are not necessarily free to pursue a same-sex relationship without consequences (as discussed later in this chapter), but they are free to have a career, to write, to have followers on social media, to feel "empowered." Yet even when they speak out, their feminism is constrained by the invisible walls of the imagined church.

Feminism in evangelical culture is not a new phenomenon. Since at least the 1980s, the debate has raged in evangelical circles about women's proper role in the church, in the marketplace, and in culture. Women have always preached, taught, and possessed authority in various ways across evangelical culture. For example, the preacher's wife has been a central figure in many churches, and in the era of the multimedia megachurch she has often become a celebrity in her own right.[7] Beyond those women connected to popular male preachers, however, there are many ordinary Christian women who have spoken up against complementarianism, and evangelical parachurch organizations have arisen promoting egalitarianism.[8] And, Evangelicalism as a whole has unevenly and slowly modernized in its view of women in the workplace.

The popular parochial feminism of the Christian influencer, however, is different than those that have come before because it is fueled by the affordances and ubiquity of digital habitus. Sarah Banet-Weiser (2018) has theorized what she calls "popular feminism," an expressive form of female empowerment that relies on and is born from digital platforms and platform capitalism. Banet-Weiser writes that "popular feminism tinkers on the surface, embracing a palatable feminism, encouraging individual girls and women to just *be* empowered" (p. 21). The politics of popular feminism, she explains, rests on the idea that visibility is political in and of itself and furthermore that it is powerful enough to preclude the on-the-ground work of activism that past feminists have done and called for. This is a feminism that is born from the algorithms that dictate our digital habitus. For example, Instagram encourages users to post photos, and the more interesting or authentic the photos are, the more likes users garner. Because of this, *showing* ourselves in an authentic way on social media becomes an oblique political act that is also part of the business plan of companies like Instagram and Facebook, whose bottom line rests on their users' disclosure.

The way that Savannah LaBrant and Rachel Hollis have presented their version of Christian womanhood has made them charismatic authorities, and has made them famous and wealthy. I define the performance of the Christian influencer by its presentation of good-Southern-girl parochialism, which includes an emphasis on their own girlishness (pink! manicures! giant smiles!), an insistence on the centrality of the family and of their role in it (usually as the matriarch), a Christian confessional style, and the performance of a cosmopolitan class habitus that sets their lifestyle apart from or tantalizingly out of reach of the majority of their audience. In other words, they perform the role of a good Christian girl who has done well for herself—who travels, who may have famous friends, who may have a fancy home—but who remains connected to her family, her roots, and above all, her faith.

Hollis, for example, is unabashed about her Christian upbringing. As she explains it, her childhood was defined by her upbringing in a small "Southern-minded" town near Bakersfield, California, where her father was a pastor in a Pentecostal church. Setting the scene, Hollis writes: "I grew up in the country. I got a shotgun for my thirteenth birthday" (2018, p. 28). Here she presents her connection to the assumed parochial background of the "good Christian girl." In other words, the type of girl who was raised to understand, respect, and occupy traditional gender roles. This is the type of girl who dressed modestly as an unmarried woman but may now be someone's "smokin' hot wife."[9] She's not a prude, nor is she sex-positive. Her sexuality is for her husband and her husband only, although she may sometimes post pictures of herself in a bikini on Instagram.

References to "Southern" values also index a conservative worldview that is part of the evangelical cultural milieu. Though Hollis was born and raised in California and spent much of her young adult life in Los Angeles, she *identifies* as Southern and parochial, which connects to an ethos that—among other things—takes a perceived natural order of gender politics as a given. Yet although Hollis is a good Christian girl, she is also a businesswoman, and much of her writing, speaking, and social media presence deals with this disconnect between embodying Christian womanhood and motherhood and pursuing a successful career.

The family is central and is always on display in the social media posts of Christian influencers. They perform and glorify their roles as mothers, grandmothers, foster parents. Popular parochial feminism centers motherhood but also allows mothers to be messy—influencers can joke about how they are not Donna Reed or Martha Stewart, but their children are always *blessings*. This connects to the confessional mode of evangelical expression. As Michael P. Young (2006) has discussed, the confessional mode stresses "a transformative religious experience" (p. 87) and provided the basis for the revivalism of the Second Great Awakening and the associated social movements of temperance and antislavery that united Protestants across the country. Having sat through many evangelical services, I can attest to the prevalence of this rhetorical mode. It is commonplace to hear Christians give their "testimony" telling their story in which they recount the worst moments of their life—for example, the experience of an alcoholic hitting rock bottom—in a packed church. These stories are powerful because they seem to provide proof of the power of the Christian faith and the personal spiritual guidance of Jesus Christ. The confessional mode is therefore a recurring performance that defines evangelical liturgy. Christians connect with the narratives of people who have failed or sinned in the past but who have been redeemed through their faith because they believe that all people struggle with sin, so to admit as much is comforting: it is authentic.

But, despite struggles in her past, despite her connection to parochialism, the Christian influencer presents herself as fitting within cosmopolitan frames. Christian influencers travel, they often live in cities, they have beautiful, well-designed homes and magazine-worthy taste. Indeed, the Christian influencer has to display good taste because this is how they sell themselves to brands and in turn how they are used by brands to sell products. Within the visual expression of a cosmopolitan lifestyle is an implicit and sometimes explicit endorsement of popular female empowerment discourses. Christian influencers are Christians first, as they say over and over again. And if they call themselves "feminist" at all, they place that second, or farther down the line of their priorities. Their feminism takes issue with the traditional role of women and wants to extend it into the marketplace, but not to destroy it. Theirs is a feminism rooted in the supremacy of motherhood: the assumed oppressed woman is a mother struggling with her choices—for example, the choice to go back to work or to become a stay-at-home mom.

This version of feminism has become a modernizing force in evangelical culture, although from the eyes of an outsider it may not seem as potent as more prominent feminist movements. The feminism of Christian influencers exists because of the ubiquity of social media (indeed, it could not exist without it), and because of that their feminism is also constrained by the medium. Like Banet-Weiser's (2018) understanding of popular feminism, popular parochial feminism is visual, it presents a vision of womanhood—usually on social media but also in books, podcasts, and conferences—that is endemic to the Christian media landscape and that normalizes an empowered Christian femininity—but what kind of empowerment becomes popular?

In the fall of 2020, while many Americans were living under lockdowns imposed by local governments as a way to curb the spread of the COVID-19 pandemic, Rachel Hollis posted a photo from the Waldorf Astoria in Beverly Hills, California. She wrote: "Dropped into LA for just long enough to get these roots in line. Now it's back home to my babes" (Hollis, 2020). Since she had announced her separation from her husband, Hollis had been jetting from her second home in Hawaii to her home in Texas. Her fans on Instagram did not respond well to this picture. They noted that they were dyeing their own hair and avoiding travel because of the pandemic. They called Hollis's behavior "selfish" and criticized her for flaunting her wealth at a time when so many people were struggling financially. Several commenters said they had lost respect for Hollis and stated that they sided with her husband, Dave, who was, as one commenter put it, "killin it at being both Mom and Dad!"

The anger directed at Hollis shows how her post displayed attitudes that fell outside of the boundaries of acceptability for popular parochial feminism. While influencers put their successful lives on display, they also have to be humble. Flaunting excessive wealth without any compunction goes against the assumption of good-girl parochial humility. Commenters also noted that Hollis was neglecting her children in favor of herself. In other words, she was not centering her role as a wife and mother. This was not accepted by her followers, many of whom went as far as to speculate that this was the kind of behavior that brought about Hollis's divorce or that her husband was right to divorce her.

There are shorter horizons for popular parochial feminism than for other feminist modes. In fact, popular parochial feminism mostly serves evangelical gender and sexual norms. Yet, although the messages may seem watered down to be simply about "girl power," within the conservative subculture of evangelicalism, that is already powerful. The popular parochial feminism of Christian influencers like Rachel Hollis and Savannah LaBrant is an entry point for many women into rethinking their own voice and place in their communities. Perhaps what proves that popular parochial feminism has potency more than anything else is that there has been a powerful backlash against Christian women influencers. In the case of the conservative subculture of evangelical Christianity, any feminism at all is surprising and, to many, threatening. And that some evangelical influencers have used their charismatic authority to lead their flock further from the norms of Christian culture in ways that threaten the shibboleths of evangelicalism (including the political conservatism that has come to define the movement since the 1980s) is deeply sinister to many Christian leaders who consider themselves guardians of traditional "family values." Because of this, many traditionally-minded Christians will tell you that these women are dangerous, heretical, and maybe even possessed by the devil.

JEN HATMAKER: TRESPASSING THE BOUNDARIES OF CHRISTIAN FEMININITY

Jen Hatmaker dedicated her 2017 *New York Times* best seller *Of Mess and Moxie* "to the girls," in reference to a Martina McBride song. She begins her book:

> This is for all the girls. The ones who thought they'd be married by now but are still single, who thought they'd be mothers by now but aren't, who said they didn't want children and have four. The ones whose marriages didn't work, the ones who found love a second time. This is for the girls who are passionate, bold, assertive; those who are gentle, quiet, impossibly dear. It's

for the decorated career slayer, the creative artist, the mom raising littles, the student in a dorm, the grandmother beginning a new venture. This is for the first absolutely living their passions and those who want to desperately but feel stuck. This one is for church girls, party girls, good girls, wild girls (I am all four). This is for all of us. (Hatmaker, 2017a, p. xviii)

Here, Hatmaker includes "all of us" as in all of us girls. Yet, as she imagines what she believes to be an inclusive audience of women readers, she also draws a picture of her version of femininity and feminine life, highlighting the type of woman she assumes to be her audience. She highlights "girls" who are not perfect, but who love Jesus. These are good parochial, Christian girls who want to be mothers but also have their own passions. Hatmaker's litany of "girls" describes the norms of popular parochial feminism. These are messy women who might not have it all together, who might struggle, but who nonetheless expect and are expected to find their identity in normative female roles such as mother and grandmother.

This "for the girls" version of popular parochial feminism is Hatmaker's brand. On the "about me" section of her website, Hatmaker first emphasizes her role as a mother and a wife and then says, "I love Jesus. I am absolutely that girl. I feel so tender toward Him that sometimes I think I'll die" (2017b). Hatmaker's emotional, indeed girly, characterization of her spirituality is part and parcel of her authenticity. She does not portray herself as a theologian or even as a woman but rather as a down-home "girl," whose love of Jesus and her family guides everything she does. In her official photograph Hatmaker is wearing large, feather earrings and smiling widely. And this earthly, fun, Kansas-raised, Texan pastor's wife has gained millions of adoring fans because she's so relatably—in her words—"messy."

Yet in April of 2016, Jen Hatmaker became a lightning rod. The event that set off her descent into heresy (if you ask some Christians) or her journey to speak her truth (if you ask others) was when the popular author, pastor's wife, and HGTV star took to Facebook to post about her support for LGBT Christians. In this post she wrote about her belief that LGBT Christians should have a place in the church, and she directed her words to LGBTQ Christians: "There is nothing 'wrong with you,' or in any case, nothing more right or wrong than any of us, which is to say we are all hopelessly screwed up but Jesus still loves us beyond all reason and lives to make us all new, restored, whole. Yay for Jesus!" (Hatmaker, 2016). Though it is written in Hatmaker's goofy, fun style, the post ignited a controversy that still rages years later. After her comments on LGBT Christians, the largest distributor of Christian books, Lifeway Christian Resources (the publishing wing of the SBC), stopped selling Hatmaker's popular books. Four years later, when you search her name on Lifeway's site, you are directed to books by other Christian writers, like Savannah LaBrant. Suddenly, after years of popularity, best-selling books, and speaking tours, Hatmaker became a pariah.

That is, in some circles. In other parts of evangelical culture she was welcomed. For example, Rachel Hollis's dedication in *Girl, Wash Your Face* reads: "For Jen, who has shaken my worldview off its axis three times: once with *Interrupted*, once with a trip to Ethiopia, and lastly by teaching us all that a *real* leader speaks the truth, even to her own detriment" (Hollis, 2018, emphasis in the original). For Hollis and empowered Christian women like her, Hatmaker was a hero and a leader who spoke out against the homophobic and patriarchal discourses that have defined mainstream evangelical culture—even when it came at a personal cost. Hatmaker added fuel to the fire during the 2016 presidential election season when she came out as one of a handful of popular evangelical NeverTrumpers. She publicly criticized presidential candidate Donald Trump for his sexist comments, exclusionary policies, and racist rhetoric and continued to beat the drum against Trump even after he won the presidency with 81 percent of the white evangelical vote and other NeverTrumpers began to walk back some of their criticism.[10]

Hatmaker continued to progress in her "affirming" theology—that is, a theology that accepts same-sex marriage as valid—and criticisms of her intensified, partly because many Christians felt that they had been blindsided by the 2015 US Supreme Court decision legalizing the practice. It has now become mainstream in evangelical circles to admit that homosexuality is not a choice but an innate orientation that cannot be "prayed away."[11] But although evangelicals express understanding for gay, lesbian, and bisexual people, they see it as the goal of the church to help these people either stay celibate or to learn how to express their sexuality within the confines of heterosexual marriage. Same-sex attraction is natural, many evangelicals will concede, but just like all other sexual desire, it has to be brought to heel within the confines of a Godly way of life. Writing in *Christianity Today*, Rachel Gilson, a same-sex-attracted woman, explained: "Our culture sings that we're 'born this way,' as if that settles the matter. But I'm born again. My life has told a different story than what society expects for me and what I expected for myself, because God himself has written his own twists and turns into the narrative" (Gilson, 2020). For Gilson, her same-sex attraction was something that she could no longer indulge when she became a Christian. She married a man and went on to have a family. All people struggle with sins of sexual desire, she wrote; people in heterosexual relationships have sexual desires for people other than their spouse, and yet they learn to overcome this to make their monogamous relationships succeed. So too should gay Christians overcome their own impulses.

Another popular Christian author and influencer, Jackie Hill Perry, who boasts nearly half a million followers on Instagram, presents another case of the limits of popular parochial feminism as a critique of evangelical gender roles. In her book *Gay Girl, Good God* (2018), Perry spoke openly about her previous same-sex relationships. But she likened her desires to Eve's forbidden desire for the Apple. These are desires warped by sin, she explained. She wrote that she knew she was gay or, as she put it, "same-sex attracted (SSA)" when she was a young child. In her teens she had girlfriends and became part of a gay community. But at nineteen Perry had a conversion moment that led her to realize that she could no longer act on her desires, even though she understood them as natural. "A common lie thrown far and wide," she wrote, "is that if salvation has truly come to someone who is same-sex attracted, then those attractions should immediately vanish. To be cleansed by Jesus, they presume, is to be immune to the enticement of sin. This we know not to be true because of Jesus" (Perry, 2018, p. 87).

Perry advises "SSA Christians" to practice self-discipline and endurance whether they decide to remain single and celibate or, like her, to pursue an opposite-sex marriage. She models the stance that has become mainstream in evangelical culture: it is okay to be gay, but ultimately, no matter who you are attracted to, your sexuality should reflect self-control rather than self-surrender. This attitude is colloquially expressed by the oft-repeated phrase: "love the sinner, hate the sin." Many evangelicals believe that gay Christians don't choose to be gay, but they can and should choose to be celibate, or, like Gilson and Perry, they can choose to marry an opposite-sex partner and have children.

Perry was one of the signatories of The Nashville Statement in 2017 (Council on Biblical Manhood and Womanhood [CMBW], 2017), which was signed by more than twenty-four thousand evangelicals including such prominent evangelical leaders as Russell Moore, James Dobson, and Richard Land. The statement expresses an evangelical view of homosexuality in the wake of the Supreme Court decision that legalized same-sex marriage in 2015. The First Article of the statement clearly and forcefully denies that same-sex marriage is biblical or acceptable. Instead, it proclaims: "We affirm that people who experience sexual attraction for the same sex may live a rich and fruitful life pleasing to God through faith in Jesus Christ, as they, like all Christians, walk in purity of life" (CBMW, 2017). In this way the statement expresses what Gilson and Perry advocate as the path for homosexual Christians—self-denial. The statement also comes out forcefully against transgender orientation. The Third Article states: "We deny that physical anomalies or psychological conditions nullify the God-appointed link between biological sex and self-conception as male or female" (CBMW, 2017). Conservative Christians often present transgender rights as the ultimate slap in the face to traditional gender roles. Many Christian commentators are openly transphobic and have made their mark in evangelical media circles by denouncing "the transgender agenda."[12]

But not only does The Nashville Statement clearly delineate a conservative Christian framework for understanding of homosexuality and transgenderism, it goes further, condemning those who would approve of LGBTQ lifestyles. The statement reads: "We affirm that it is sinful to approve of homosexual immorality or transgenderism and that such approval constitutes an essential departure from Christian faithfulness and witness" (CBMW, 2017). Coming on the heels of Hatmaker's public acceptance of same sex marriage, this reads like an attempt to formally rebuke her charismatic authority and others like her who would preach outside of the church walls. Hatmaker tweeted her disdain for the statement after it was announced, writing: "The fruit of the 'Nashville Statement' is suffering, rejection, shame, and despair. The timing is callous beyond words" (Hatmaker, 2017c). Hatmaker's words were covered as part of the news of the statement in major media outlets including *USA Today* and the *Washington Post*. Hatmaker was content to provide the seemingly lone voice for affirming theology against the official words of such established figures in Christian culture as John Piper, James Dobson, and Russell Moore.

Despite the clear and powerful backlash to Hatmaker's affirming theology that followed her initial statements on same-sex marriage, she has not retreated. She has advanced. In June of 2020, Hatmaker's young adult daughter, Sydney, publicly came out as gay on Hatmaker's popular *For the Love* podcast. In a tearful interview with her daughter, Hatmaker talks about how her affirming theology developed. Sydney explains the struggle she went through as a "Jesus freak" kid who in early adolescence began to realize she was gay. Hatmaker calls her slow journey to affirming theology as "one of her greatest sadnesses" because she believes it hurt her young daughter while Sydney was beginning to understand her identity. In this same podcast episode Sydney encourages Christians to protect transgender people, especially trans children. This is particularly important, Sydney explains, because "especially in affirming Christian spaces, this readiness to talk about gay people because we understand them more, and see them more represented in our lives, and can digest them easier—and not want to talk about trans issues" (Hatmaker, 2020).

Hatmaker ended this episode with an entreaty addressed to Christians: "I want us to have a reckoning together that when we refuse to cherish and affirm the LGBTQ community, including our kids, we are literally breaking their hearts. We are breaking their bodies. We are breaking their minds. This is not neutral. This is not just a difference of opinion. This is causing harm and trauma and suffering" (Hatmaker, 2020). And as with every controversial event in the Hatmaker canon, the backlash to this podcast episode was swift and vitriolic. One response put it bluntly:

> It is now abundantly clear why Jen Hatmaker has come out gay-affirming— she hates her daughter. In fact, she hates her daughter so much that instead of being willing to lay down her own reputation, her own pride, and her own self-satisfaction to see her daughter have an opportunity to hear the truth of the gospel and respond to it, she instead lies to her daughter and gives her false hope while affirming her affront to God. (Maples, 2020)

Another online outlet ran with the headline "Apostate Author Jen Hat-maker Reveals Her Daughter Is Lesbian 'in Honor of Pride Month'" (Clark, 2020). But as many Christian blogs and news outlets printed their vitriolic dissent, the social media sphere responded differently. When Hatmaker posted a link to this podcast on her Twitter, many Christians weighed in with their own experiences. One wrote: "And now: I feel less alone. So does my son. Who's mom is also a christian 'leader' of sorts... well done sisters. And Sydney? Your mom's people are your people. Welcome to the crazy tribe."

With what traditional evangelicals see as her contrarian political stances, Hatmaker has trespassed the boundaries of evangelical feminin-ity and become a liability to those evangelicals invested in maintaining the cultural boundaries that define this conservative culture. And pow-erful institutions and influential people have tried to silence Hatmaker in a variety of ways: through public shaming, by denying her a space in Christian bookstores, and through official theological statements that obliquely rebuke her. Jen Hatmaker's case shows how popular parochial feminism speaks in the vernacular of evangelicalism but sometimes uses this vernacular to express ideas that make more conservative Christians bristle. This is what makes Hatmaker such a divisive figure. Not only is she unabashedly liberal-minded, she has retained her charismatic authority despite consistent backlash from prominent evangelicals—many of whom occupy formal positions of authority. Much of the criticism of Hatmaker springs from the fear that she is radicalizing her audience of evangelical women with what many see as her left-wing political opinions. To con-servative evangelicals, Hatmaker is a wolf in bright pink sheep's cloth-ing. Ultimately, empowered, outspoken women like her are perceived as dangerous by the evangelical power structure because their charismatic authority allows them to circumvent traditional power dynamics.

HASHTAG ACTIVISM: FEMINIST COUNTERPUBLICS ON SOCIAL MEDIA

On April 27, 2017, Rachel Held Evans tweeted in all caps: "LITERALLY THREW MY PHONE ACROSS THE BEDROOM OVER THIS PIECE." She was referring to a *Christianity Today* article titled "Who is in Charge of the Christian Blogosphere?" written by a female Anglican priest, Tish Harrison Warren (2017). According to Warren's article, popular women Christian bloggers were causing a "crisis" in evangelical culture, and the chief offender was Jen Hatmaker. For Warren, Hatmaker was a symptom of a larger problem—namely the fact that many popular Christian women bloggers have massive followings on social media and thus outsized influence. As a corrective, Warren's article called for increased accountability in, and institutional oversight of, the Christian blogosphere. It was not clear what exactly this might mean, but the article seemed to suggest that various denominational authorities should be somehow regulating or editing Hatmaker and other influencers—and bringing them in line.[13]

Rachel Held Evans's was not the only impassioned response to this article. The social media reaction centered around Twitter in the days after the article's release but also saw responses cropping up on Medium, personal blogs, and Facebook. Hundreds of women recounted personal stories of being silenced or backgrounded in evangelical churches and Christian communities. For many of these women the Christian blogosphere and social media have provided a means to get around the patriarchal structure of church culture. Because of this, on and around the hashtag created by *Christianity Today*, #AmplifyWomen, Twitter users passionately defended the role of the female charismatic authorities in evangelical culture.

Watching the tweets come in with the hashtag #AmplifyWomen, and seeing them quickly retweeted by Evans, Hatmaker, and other prominent Christian women, I saw a movement of Christian women erupting onto the public sphere. Many women who once may have remained silent, having no outlet and no support for their opinions, were speaking up on social media against the idea that their role models should be subject to formal authority. Editors at *Christianity Today* responded to the robust and diffuse discussion that followed the article's release by adding a note to the original piece that in part stated: "The conversation continues to spread and split into what scientists call a dendritic—a series of branching pathways that resemble a tree or a nervous system" (Warren, 2017). This "dendritic" maps the beginning of a feminist movement in evangelical culture that is opening new spaces, asking new questions, and raising new issues. Even though this movement centers around popular parochial

THE INFLUENCERS   119

feminism, it stretches and changes this ethos as women connect to others and form inchoate feminist networks. The #AmplifyWomen conversation was a moment in which the contours of a new counterpublic made up of evangelical women could be seen. And this counterpublic would go on to change evangelical culture in the #MeToo movement of 2017 and 2018. Yet the #AmplifyWomen conversation is worth analyzing in itself as it is the result of a perceived attack on an evangelical influencer, Jen Hatmaker, and as such shows the potency of popular parochial feminism as a change agent in evangelical culture.

The charismatic leaders at the center of this movement—Jen Hatmaker, the target of Warren's article; Rachel Held Evans, who shepherded much of the conversation on social media; Beth Moore, another popular #NeverTrump Christian; and other women who posted on their own blogs and social media accounts about the article—were contributing to a counterpublic of Christian women who have used the affordances of social media to connect with others and create relationships with like-minded women. Michael Warner (2002) has explained that counterpublics are those publics that go against the grain of the norms of a larger public, in this case the public of evangelical culture. Counterpublics imagine the world differently than the normative public. Important for Warner is the fact that counterpublics are not only activist in the traditional political sense, but in the social sense; they want to change the way that social life works (2002, p. 122). What has become known as "hashtag activism" has been a particularly fruitful practice that allows for feminist counterpublics to form. As Sarah J. Jackson, Moya Bailey, and Brooke Foucault-Welles (2020) wrote in the their analysis of feminist Twitter, these networks "become spaces where a growing number of people, connected by their use of hashtags and the shared trauma that inspired their deployment, can amplify the same kinds of feminist critiques that have often had only limited or elite reach" (p. 3). Christian women have always used mediation to find audiences outside of the church walls—from printed tracts, to radio shows, to television.[14] Yet with social media, ordinary women can more easily contribute and grassroots movements can coalesce and grow.

What erupted through the #AmplifyWomen hashtag was a counterpublic that has been many years in the making, one uneasily defined by both contemporary feminism and evangelical theology.[15] The women who shared their experiences and opinions on the hashtag expressed a collective desire to change the evangelicalism that they had grown up with— a culture that they characterized as oppressive to women. The (mostly) women and (some) men who engaged with the debate that flurried around the *Christianity Today* article deployed various cultural touchstones, often displayed a fluency with the argot of academic feminism and contemporary pop-culture postfeminism[16] as well as with theological justifications for their understanding of their own role and the role of women in evangelical culture and society. The stories women told on this hashtag shed light on how ordinary evangelical women understand social media and social media celebrities. As such, this hashtag is a useful site to understand the character of the evangelical feminist counterpublic.

Explaining why she disagreed with Warren, Christian writer Keri Wyatt Kent echoed a sentiment that was often repeated in the #AmplifyWomen conversation. She wrote on her blog: "To ask why women don't serve within the authority and structure of the church is a bit tone deaf. The people 'in charge' of many Christian women have told them to sit down, be quiet, or go change some diapers in the nursery" (Kent, 2017). For Kent and others, social media platforms have become places of refuge for Christian women, much like the ministries from which women have traditionally exercised their authority. But unlike those ministries, the internet has provided women with platforms from which they can preach to large audiences, albeit unofficially. Digital habitus has liberated and amplified many female voices, and because of this, to corral influencers under the authority of church culture, as Tish Harrison Warren seemed to suggest was the purpose of her article, would be akin to telling women to go back, sit down, or to change diapers in the church nursery.

For the women who used the hashtag to express their frustration with Warren's article and the attitudes of the conservative Christians it seemed to represent, blogs and social media provide a corrective to the entrenched patriarchal authority of the church world. As such, many argued, bloggers and posters should not answer to any authoritative structures. A Twitter user wrote: "If I thought the broader church was completely right about everything (especially women in ministry), I WOULDN'T HAVE A BLOG" (emphasis in the original). Another wrote: "Hey, institution that could be leading ppl astray. Would you ensure I don't say anything you think would lead ppl astray?" Here, these women cast the blogosphere and social media as workarounds, ways for women to speak without having to submit themselves to the authority of a church that, as another user pointed out, "has almost exclusively been shaped by straight white men."

As a corrective to a culture seemingly controlled by white men, then, women have taken to the internet, to their own blogs and social media pages. In doing so, they have provided leadership and witness for other women. As one user tweeted, "The blogosphere gave me the female leaders I was looking for, but couldn't find in the Church—women standing up for justice." The internet allowed women like this access to a community they might not have otherwise found—a community of like-minded women thinkers who were willing to talk about discrimination, sexism, and abuse in church culture. Warren's piece, which seemed to call the authorities on unordained bloggers, bothered the women who posted on the hashtag who fought to keep social media a protected space for women's voices.

Popular writer Ann Voskamp[17] wrote a response on her blog in a litany style that emphasized her humility as an unordained woman and worked to connect that humility to other biblical figures. "Yeah, I don't know much of anything," she wrote, "except that we all need each other, that we all belong to each other, but seems like maybe God has always chosen women who felt less than, women that no one thought were enough: Tamar was harlot, Ruth was an outsider, the wife of Uriah , who became the wife of David, was an adulteress, and Rahab was a woman of the night" (Voskamp, 2017). This defense and the style in which it is voiced speaks to evangelicalism as a populist, democratic religion where humility and humble beginnings are often prized over institutional authority.

Evangelicalism tends to believe itself to be a folk religion and is littered with figures, who, just as these women did, used their own biblical exegeses as a means to establish their authority. Since the Second Great Awakening antiauthoritarian and anticlerical sentiments have been prevalent, and charisma and popularity have been prerequisites for leadership (see Fitzgerald, 2017, pp. 13–48). Mark Noll (1994), in exploring the anti-intellectual "desires" of evangelicalism, has noted that "the evangelical ethos is activistic, populist, pragmatic, and utilitarian" (p. 4). Because of this, the idea of evangelicalism as a populist religion marries well with the idea of the internet as a democratic space.[18] Blogging is seen as an inherently democratic medium—a common understanding that many users pointed out. For example, one wrote: "'Some of these bloggers aren't accountable to formal church authority!' That's not a bug, it's a feature." And another: "The fact that churches can't police... diverse voices on the internet is.... what makes the internet great." These tweets portray Warren and evangelical authorities who might police the internet as behind the times, clinging to a structure that the internet has blown up. They seem to suggest that evangelicals are struggling with the democratization of information itself. Austin Channing Brown (2017) addressed this with recourse to the history of media use in evangelical culture and the moral panics that have accompanied technological change. She tweeted: "We survived the printing press, radio, televangelists.... I think we will survive the blogosphere, and whatever is next. I'm not worried."

In addition to their spirited defense of new media's affordances, the public that sprang forth around the #AmplifyWomen hashtag expressed ideas that ran counter to the perceived image of conservative evangelicalism. Like Hatmaker, these women do not always accept evangelical cultural or political orthodoxy, and the passionate nature of the public that rose up to defend Christian influencers shows that ordinary women are listening to the women who populate their Twitter and Instagram feeds. On the #AmplifyWomen hashtag, women explained why they responded to female bloggers and influencers and why they would not concede the connective power digital habitus offered them. In the months and years that followed, Christian women mobilized the connections they made to enact real changes in evangelical culture and power structures.

BETH MOORE AND THE #METOO MOVEMENT

In October of 2017 the actress Alyssa Milano tweeted: "If you've been sexually harassed or assaulted write 'me too' as a reply to this tweet" (Milano, 2017). Thus she ignited what has come to be known as the #MeToo movement. Since then, this movement, which had begun as a campaign to support victims of sexual assault by Tarana Burke a decade earlier, took on a life of its own as it entered into public consciousness on social media.

The #MeToo movement has mobilized large-scale campaigns that have toppled famous abusers in media industries and in politics. It has also forced a parallel reckoning in evangelical culture that was fueled by Christian influencers and the networks of engaged women who follow them. In fact, on the day that Milano tweeted her entreaty to women, two unlikely replies came from prominent evangelical women. Kay Warren, the wife of celebrity pastor Rick Warren, and Beth Moore, who became a de facto leader in the Christian #MeToo campaign.

Moore started her ministry career teaching exercise classes in her church as a way to teach and preach in a denomination (the SBC) that does not allow women to be ordained. Later, she established her own brand of video Bible studies and eventually the organization Living Proof Ministries. Moore headlines Christian conferences and has nearly one million followers on Twitter. She fits the popular parochial feminist mold but as a representative of the older generation of evangelicals (Moore was born in 1957 and is a grandmother), she brings a gravitas and an understanding of the difficulties of achieving empowered femininity in evangelicalism to her role. In a profile of Moore in *Texas Monthly*'s "Power Issue," one journalist captured Moore's popularity:

> Petite and blond, Moore dresses in expensively casual outfits. She jokes about her big hair and her copious use of self-tanner. She talks self-deprecatingly about her domestic life. Then she grabs a Bible and flips through the pages with perfectly manicured fingernails. "Anybody need a fresh dose of Jesus?" she asks. Her fans scream some more as they take photos of Moore with their cellphones. (Hollandsworth, 2018)

Because Moore is a member of the SBC, her rise has been all the more difficult given the organization's complementarian stance. In the SBC women are typically not allowed to preach, and even as Moore herself identifies as a complementation, she became a type of preacher by becoming a media celebrity. And like the social media influencers analyzed earlier in this chapter, Moore presents herself in the frame of popular parochial feminism. The "About Beth" section of the website of Living Proof Ministries notes that "at the age of 18, Beth sensed God calling her to work for Him. Although she couldn't imagine what that would mean, she made it her goal to say yes to whatever He asked" (Moore, 2018a). Here, her theological bent is characterized as coming from a place of girlish passion and her pursuits, though deeply felt, as generally submissive. As Emma Green's profile of Moore in *The Atlantic Magazine* revealed, Moore has used this tactic throughout her career. Green (2018) wrote about how Moore "spent her career carefully mapping the boundaries of acceptability for female evangelical leaders."

Like Hatmaker, Moore was vocal about her disdain for candidate Trump in 2016 and remained a critic of his presidency. And in 2018, at the height of the #MeToo movement, Moore penned an open letter about the sexism she faced as a woman in the SBC. She wrote:

> As a woman leader in the conservative Evangelical world, I learned early to show constant pronounced deference—not just proper respect which I was glad to show—to male leaders and, when placed in situations to serve along-side them, to do so apologetically. I issued disclaimers ad nauseam. I wore flats instead of heels when I knew I'd be serving alongside a man of shorter stature so I wouldn't be taller than he. (Moore, 2018b)

Moore explained how the sexist culture of the SBC constrained her. She was offered a seat at the table, but only if she would act in a submissive and nonthreatening manner.

As she spoke out against sexism in the SBC, Moore also led the charge on social media for women and men to discuss sexual abuse in and out-side of the church. Moore was abused as a child and states that the church is where she was able to find refuge, but she has opened up a space for women on social media to discuss the abuses they have faced within churches on the related hashtags #ChurchToo and #SBCToo. At a confer-ence headlined by Moore in 2018 popular preacher, writer, and evangeli-cal media celebrity Max Lucado revealed that he too had been abused by a member of his church community when he was a child (Lee, 2018). It is interesting to note that this conference took place at the Billy Graham Center at Wheaton College, as Graham famously did not allow himself to be alone in a room with any woman who was not his wife. This has become known as "the Billy Graham rule," which many male evangelical pastors have purported to follow. In the days of Billy Graham's dominance in Christian culture, frank discussions about sexual abuse like the one that Beth Moore led in 2018 would be unheard of.

The head of the SBC at the time, Al Mohler, wrote on his website: "The last few weeks have been excruciating for the Southern Baptist Convention and for the larger evangelical movement. It is as if bombs are dropping and God alone knows how many will fall and where they will land. America's largest evangelical denomination has been in the headlines day after day. The SBC is in the midst of its own horrifying #MeToo moment" (Mohler, 2018). Mohler indicated the power that the #MeToo campaign had on the reputation and structure of the SBC. And the revelations of sexual abuse and reconsiderations of complementarianism theology defined the conversation at the annual meeting of the SBC in 2018.

But the discussion was far from over; there were more bombs yet to fall. In 2019 the *Houston Chronicle* put out a six-part series detailing allegations of sexual abuse by pastors and church leaders in the SBC. Outrage on social media followed. During this time Beth Moore posted a picture of herself as a young girl alongside the tweet: "We understand how you feel. We didn't want to know about sexual abuse either" (Moore, 2019). Other Twitter users replied with their own pictures from their childhood, showing the age when they were sexually abused, and still others quote-tweeted Moore with their own stories. In this way Moore facilitated a public reckoning about abuse in the SBC that weaved stories and narratives from ordinary survivors.

These movements had an earlier antecedent, in the discourse surrounding the two hashtags #ThingsOnlyChristianWomenHear and #Things OnlyBlackChristianWomenHear—both in 2017. The former hashtag was started by Christian writer Sarah Bessey, who wrote *Jesus Feminist* (2013). This hashtag saw women sharing stories of being shamed and in some cases abused in and around church culture. A typical tweet that used this hashtag referenced the subtle sexism implicit in Christian culture as women recalled being told things like "You better cover up or you'll make boys sin." Other users shared stories of abuse and oppression, like the woman who wrote: "My pastor told me to be a good wife and my husband wouldn't beat, rape and try to kill me." This hashtag became a discursive space in which women shared their stories of abuse and pushed back against people who tried to discredit their experiences.

The hashtag #ThingsOnlyBlackChristiansHear was started in response to #ThingsOnlyChristianWomenHear by Ekemini Uwan, a popular podcaster and theologian, as a way to indicate that the initial hashtag hailed a public that was defined by and began with the experience of white

Christians. This hashtag dealt not only with gender discrimination but also with racism in evangelical culture and with the intersectional subjectivity of Black Christian women. For example, one user recalled hearing in church: "'It wasn't wrong for Christians to have slaves. They had slaves in the Bible.'" As the online magazine *Faithfully*, a publication that focuses on the concerns of Christian women of color, explained, the hashtag #ThingsOnlyBlackChristianWomenHear did not garner the same amount of public attention outside of the Twitter discussion as the initial hashtag, which was covered by several Christian media outlets (see Menzie, 2017). These dueling hashtags reveal another partition in evangelical culture between white Christians and Black Christians (discussed in more detail in chapter 5), and the lukewarm response that #ThingsOnlyBlackChristianWomenHear received in comparison to other hashtags show that the evangelical power structure values the concerns of white women more than those of nonwhite women.[19]

Despite the fact that they were regarded unevenly, like the #Amplify Women conversation, these hashtags provided a space for a network of Christian women to form. And in these cases, women discussed abuse across church culture. This opened the floodgates for the #MeToo movement to enter into the evangelical sphere. Moore's open embrace of the #MeToo movement and her ability to retarget it toward Christian spaces, especially the authoritative structures of the SBC, made her controversial. Many conservatives Christians—most of them men—asked, Why were so many people listening to Beth Moore? Women, after all, were not meant to preach and indeed are not allowed to be ordained in her denomination.

To her critics, Moore represented the threat of feminism that has been a foil to evangelical leaders since the 1970s, but this time the call was coming from inside the house. At a public event, John MacArthur, a popular pastor and writer, lambasted Moore as a huckster who was only fit to sell jewelry on QVC and said she should "go home" (Relevant, 2019), implying that Moore's proper place was in the domestic sphere. And another Baptist pastor, Josh Buice (2019), called for the SBC, including Lifeway, to cut all ties with Beth Moore—the same disciplinary measure previously used against Jen Hatmaker. Criticisms like these ignited fires on social media, where popular Christian women and others rose up to defend Moore.

Ultimately, evangelicalism could not outrun the #MeToo movement. In 2018, Bill Hybels, the leader of Willow Creek Community Church, one of the most prominent megachurches in the country, resigned after allegations that he had sexually harassed women employees. The case against Hybels involved the reports of multiple women and dated back to encounters in the 1990s. It was alleged that for at least five years other prominent evangelicals had successfully shielded Hybels from repercussions.[20] And, in Moore's home denomination, the SBC, the president of the Southwestern Baptist Theological Seminary, Paige Patterson, was fired after reports of inappropriate behavior surfaced. The ripples of the #MeToo movement in secular culture that had toppled Harvey Weinstein and others were being felt in evangelical culture. Although influencers like Moore ignited the conversation on social media, it was sustained by ordinary social media users. Thus, in an evangelical culture that is suffused with digital habitus, grassroots narratives can take hold relatively quickly, counterpublics can form and gain power, and as was the case with the #MeToo movement, unlikely evangelicals can change established cultural discourses.

Popular parochial feminism sometimes leans up against the complementarian dominance of evangelical culture, but it rarely tips it too far. However, there are those influencers, chief among them Jen Hatmaker and Beth Moore, who have used their charismatic authority to start and sustain difficult conversations in evangelical culture. The fact that they have remained popular and have even grown their followers, even after being publicly shamed by evangelical cultural powerhouses like Lifeway and *Christianity Today*, is troubling to those who are invested in maintaining the authority structures that have defined evangelical culture. In the #MeToo and related hashtag campaigns and through the leadership of Beth Moore, we see that these networks, these flocks following Christian influencers, inspired by their charismatic authority, have made strides and changed the way that evangelicalism works. They have used their charismatic authority to do things, to change things—and they won't be silenced.

CONCLUSION

In March 2021, Beth Moore cut ties with the Southern Baptist Convention. She announced that Lifeway Christian Resources would no longer publish her books.[21] Her departure came after she had spent nearly five years publicly pushing back against the SBC's endorsement of Christian nationalism and President Trump and the organization's failure, in Moore's eyes, to meaningfully correct and atone for the abuse and sexism in its ranks. Because of her celebrity and influence, Moore's announcement was covered widely in the popular press from CNN, to NPR, to *USA Today*. Many reporters, evangelical insiders, and observers publicly pondered whether thousands of Christian women might follow Moore's lead. As with the evangelical anxiety around any of Moore's moves, the fear was that her influence and power might lure women away from the traditions of the past.

Evangelical influencers have proven that a lot of women think like Rachel Held Evans, like Jen Hatmaker, like Beth Moore. In fact, they fill stadia all over the country, they read posts, they comment, they tweet, they start their own blogs, podcasts, social media accounts, and they feel less alone and more empowered to speak their own truths. And this is a problem for those invested in maintaining the entrenched gender politics of evangelical culture. That is why the vested authority structures of evangelical culture have fought back. Lifeway Christian Resources, the SBC, famous pastors, have all put their thumbs on the scale in hopes that their cultural power might balance out the charismatic authority of these sirens.

But digital habitus has fundamentally changed the authority structures of the evangelical church. The version of feminism embodied by evangelical influencers promoted the idea that women should have equal authority to men, they should be able to voice their opinions in public in any way that they see fit, and their theological and cultural opinions should be taken seriously. Their version of feminism does not include sexual liberation and only scratches the surface of gender as a construct, and in these ways they do not resemble second-wave, third-wave, or postfeminist feminists. And yet their understanding of female empowerment indicates that there is a grassroots movement of women from a conservative subculture who identify or are beginning to identify as feminists.

This is no small thing given the public opposition that evangelicals have put up to feminism for the past forty years. And it has also gone beyond just talk—the #MeToo movement has swept through evangelical institutions, backlashes have incited counterbacklashes, and true evangelical counterpublics have grown and thrived in the social media space. The democratization of information has in some ways liberated female voices in evangelicalism. And as women have created counterpublic spaces on social media, the power of evangelical influencers has gone beyond the hashtag. It has changed the way that evangelical culture sees, hears, and sometimes listens to women.

5 Racial Reckoning and Repair

THE URGENT CONVERSATION ABOUT RACE ON THE BLACK CHRISTIAN PODCAST CIRCUIT

The footage was graphic and brutal—and it was everywhere: posted on Facebook, retweeted, broadcast in clips on the news. It was eight minutes and forty-six seconds long, and it showed a white police officer kneeling on the neck of a Black man. The man pleaded for the officer to stop, he pleaded for his life, and finally, he cried out for his mother. Videos depicting police brutality against Black people had gone viral before, but there was something about the moment that this video entered into the public consciousness that made it different.

On June 1, 2020, the *Pass the Mic* podcast went live on Facebook. In a raw, freewheeling conversation the show hosts, Jemar Tisby and Tyler Burns, talked through the collective effervescence of the inchoate uprisings sparking in every state in the nation in response to George Floyd's murder. Tisby and Burns spoke about what the uprisings might mean for the future of racial justice in the United States, but they also spoke about the Christian church in America. Tisby put it this way: "Maybe the place to start for us is revolution in the church" (Tisby & Burns, 2020a).

American evangelicalism is not Black and white. It encompasses thriving communities of Asian churches, Latino churches, and multiethnic

churches,[1] and it crosses national and cultural boundaries. However, there
is a historical and sociological narrative about evangelicalism in both
scholarly and popular discourse that does divide American Protestants
into the broad racial categories of Black and white. Christians understand
the schism between the Black church and white evangelicalism as a central
dividing line in American Christian culture even as they also understand
and acknowledge the artificiality of dividing American Protestantism in
this way. And in the media and politics, Black and white Christians are
understood to have distinct cultures and often opposing views. Thus the
distinction between Black and white versions of Christianity in America
are clear from both emic and etic perspectives. Since the 1990s, however,
there have been calls to more meaningfully unite American Protestantism
and to dismantle the barriers between these two distinct traditions.
Although "racial reconciliation"—the idea that Black and white Christians
can find ways to set aside their differences and unite the church—has been
in the evangelical lexicon for decades, this discourse has typically been
promoted by white evangelicals and has been constructed on their terms.
But in 2020 the white gatekeepers that once led the conversation were
sidelined by passionate Black media makers intent on reshaping the narrative to match the urgency of the moment.

Through interviews with podcasters and a discourse analysis of three
podcasts focusing on the Black Christian experience, I look at the racial
reckoning the evangelical church is facing as well as the people and media
that are leading the conversation on it.[2] In chapter 4 I analyzed evangelical women who have used social media to create feminist counterpublics
in evangelical culture. This chapter argues that podcasts have become
another site of counterpublic discourse, and I specifically chart how Black
Christians have used the affordances of the podcast—especially the way
that podcasters can speak to multiple audiences at once—to reconstruct
the discourses around race in evangelical culture that have historically
been led by white Christians. The discourse represented by these podcasts
is rooted in the Black Christian experience of racism in white evangelical culture, and it offers a blueprint forward for both Black and white
Christians in the age of Black Lives Matter. But it remains to be seen
whether white evangelicals are really listening.

THE RACIAL RECONCILIATION MOVEMENT

Political pollsters often divide evangelicals based on race: white evangelicals are seen as a powerful, politically conservative voting bloc, whereas nonwhite evangelicals are rarely polled or profiled as such.[3] The distinction that pollsters make between white and nonwhite Protestants reflects an actual cultural and political difference between those churches and denominations run by white evangelicals and those churches and denominations run by nonwhite Christians. But the distinction is also generative in that white evangelical individuals recognize themselves as part of this political public so often defined by pollsters, whatever their own political beliefs may be. Black Christians also recognize the assumed cultural and political norms of this public and understand the racial makeup of evangelicalism in this way.

Though I do not wish to reify the distinction between Black Christians and white Christians, especially insofar as doing so backgrounds the vibrant and important traditions of myriad other Christians, it is important to understand that evangelical culture is raced and that the central understanding of race in American evangelicalism is between Black and white Christians. There are historical reasons for this divide. Namely, that the Black church constructed itself against a white, American Christianity that excluded Black people and both explicitly and tacitly endorsed violence and discrimination against nonwhites. As James Baldwin commented: "I don't know if white Christians hate Negroes or not, but I know we have a Christian church which is white and a Christian church which is Black."[4] In his *New York Times* best-selling book *The Color of Compromise* (2019), Jemar Tisby put it in similarly blunt terms: "Harsh though it may sound, the facts of history nevertheless bear this truth: there would be no black church without racism in the white church" (p. 52). Although many white evangelicals often nostalgically and counterfactually look back to the abolitionist movement of the nineteenth century as a moment when Christians stood up on the right side of history, white Christian America at its worst was complicit in the institutions of slavery and white supremacy that defined the growth of the American republic. And even at their best, white Christians have been remarkably recalcitrant and conservative in their visions for racial justice.

Many historians, focused on the earlier days of the American republic, have outlined how Christians provided the theological justifications
for slavery and the subjugation and dehumanization of African people.[5]
Even those northern Christians who began to question slavery after the
American Revolution were generally conservative and "believed that the
more obvious abuses of the system would dissipate with the conversion
of masters and slaves" (Emerson & Smith, 2001, p. 49). It was not until
the 1830s that abolition was able to gain any traction among northern
American Christians.[6] But even during this time there was no Christian
consensus that slavery was wrong, and many southern and northern
denominational authorities continued to support the practice in explicit
and implicit ways.

Though slavery and the Civil War provide the backdrop for the creation
of what is known as "the Black church,"[7] this history has been well covered by other scholars. I want to center my conversation about the racial
schisms that continue to define American evangelical culture in two eras:
the Civil Rights movement and the Promise Keepers trend of the 1990s.
The conflicting styles and subjectivities at the heart of the Black church
and the white evangelical church developed and were on display during
both of these periods. The fault lines that they highlight remain central to
Christian race relations today.

In 1953, Martin Luther King Jr. wrote that "it is appalling that the most
segregated hour of Christian America is eleven o'clock on Sunday morning," (King Jr., 1963). referring to the racially segregated nature of church-
going in the United States during that time. Although historically Black
churches became central organizing spaces in the Civil Rights era, the fact
that white Christian communities were segregated from Black churches
made the Civil Rights movement's emphasis on racial justice seem like
a faraway problem to many white Christians. Billy Graham began integrating his audiences in the 1950s; however, his magazine, *Christianity
Today*, conspicuously ignored Civil Rights and refused to cover or discuss
the issue. Anthea Butler has characterized Graham's attitudes about race
using the term "evangelical gentility". Butler notes that Graham "recognized the problem of racial injustice and evoked the pain caused by unjust
social norms but he was unwilling to break ranks with the white status
quo" (2021, p. 44).

Although he made gestures to the importance of racial parity, Graham believed the Civil Rights movement was a threat to the American way of life both because he found the civil disobedience tactics employed by Civil Rights leaders to be disruptive and because he feared that the movement was connected to communism.[8] Graham's attitude is perhaps indicative of the body of white evangelicals for which he was a figurehead. Though there were some exceptions, white evangelicals overwhelmingly did not support efforts toward racial equality and racial justice. In fact, the fear of racial mixing often provided the impetus behind evangelical political involvement, especially in California, where evangelicals threw their support behind municipal measures that would allow for continued housing discrimination (see Dochuk, 2011, pp. 170–172). While Black Christian churches and organizations fought for Civil Rights, many white Christians organized against what they saw as radical racial policies like school integration. And others remained indifferent or tentative on the issue of race.

In addition to the marches and speeches that defined the Civil Rights movement in the news media, a new consciousness formed among Black Christians during this era. Two seminal works were published in 1969: James Cone's *Black Theology and Black Power* and the paperback edition of Howard Thurman's *Jesus and the Disinherited*, which had come out twenty years earlier. These works theologically linked two expressions of Black identity—the Christian and the empowered—and became touchstones for Black Christians.[9]

Andrew Billingsley (1999) has argued that "in times of extreme and sustained crisis, the African American community will turn to the churches and their ministers for comfort, support, leadership, and guidance" (p. 185). And in the tumult of the 1960s the church provided a cultural space for Black Christians to collectively construct a new vision for racial equity in America. Many of the central figures of the Civil Rights movement came from the church world and organized in church spaces. The culture of the Black church, including the evolution of Gospel music during the Great Migration, was intwined with the fight for Civil Rights and evolved along with it, as Nick Salvatore's (2005) biography of the preacher C. L. Franklin (father of Aretha) explores. And so the Black church developed distinct cultural, liturgical, and theological traditions that centered on the fight against racist oppression. This is often referred to as the "prophetic voice" of the Black church.

But as the activism that defined the Civil Rights movement faded in the 1970s and 1980s, the pervading feeling among white evangelicals was exasperation that race was still considered an issue—they instead preached the "colorblind gospel," which in effect made race invisible (see Dochuk, 2011, p. 274; Walton, 2009, p. 189). And even Black church culture decentered racial justice as it moved toward new foci. For example, Black televangelism, which became popular in the 1980s, represented a break with the animating philosophies of the Black church of the Civil Rights movement—which focused on community-building, organizing, and activism—and the beginning of a mediated, visual form of Black Christianity that has become a shaping force throughout the African diaspora (see Frederick, 2015, p. 38; and Walton, 2009). Paula McGee (2017) has written about the rise of the megachurch pastor and televangelical celebrity T. D. Jakes as the apotheosis of what she calls "the new Black Church," which emphasizes branding and growth more like a corporation than like the Black church of the Civil Rights movement.

Into this landscape, in the 1990s some evangelicals began to insist on a new focus on racial healing that they believed was inspired by the Civil Rights movement. Michael Emerson and Christian Smith (2000) have explained that "evangelical leaders picked up a seemingly forgotten piece of Martin Luther King's vision—the need to reconcile races—and ran with it" (p. 166). This was the basis of the racial reconciliation aspect of the Promise Keepers movement, a Christian men's movement that became popular in the mid-1990s and saw Christian men filling football stadia to sing along to Christian rock songs and partake in what they believed was a modern vision for Christian masculinity.[10]

Initially the organization faced backlash from many of their white members who thought the emphasis that leaders and speakers placed on race was counterproductive or irrelevant to celebrating Christian masculinity, but Bill McCartney, a white former football coach, pushed his message of racial reconciliation to the center of the Promise Keepers' messaging because he believed he was following a spiritual directive. McCartney soon attracted Black pastors who spoke at Promise Keepers meetings and appeared arm in arm with white Christians to symbolize the power of racial unity in the church. The vision of racial reconciliation that the Promise Keepers put into the evangelical consciousness was one that believed in the toxic nature of white supremacist thinking and behavior among white Christians but also placed responsibility for reconciliation on Black Christians who, it was assumed, needed to repent of their bitterness and their own racism against white people.[11]

White evangelicals may have been inspired by Martin Luther King Jr., but their version of racial reconciliation as popularized by the Promise Keepers shifted the focus of racial justice from systems and institutions to individuals and relationships; this vision for racial equality relied on the idea that Christian repentance could cure individuals of their racist leanings, and as people from different races interacted with each other and become friends, racism would naturally disappear. This attitude helped white evangelicals and institutions sidestep calls for broad social reform and instead allowed them to focus on improving interpersonal relationships among Black and white Christians. In addition to its refusal to confront systemic, institutional racism, the Promise Keepers movement only focused on men, which as Chanequa Walker-Barnes (2019) has explained, was a blind spot that "resulted in a single-axis understanding of racism, that is one that ignores its interactions with other forms of oppression, especially gender oppression" (p. 64). Whether the message of racial reconciliation that the Promise Keepers laid out was problematic, or watered down, it shaped how white evangelicals understood race as they entered into the twenty-first century.

In the 1990s other symbolic events seemed to confirm to white evangelicals that racism in the church was behind them and that the onus of racial reconciliation rested on individuals. In 1994 the Pentecostal/Charismatic Churches of North America (PCCNA) was formed in an event that was dubbed "the Memphis Miracle" because it saw the remerging of two denominations that had been split along racial lines, and the white members issued a public apology to the Black members. A similar admission of guilt happened in 1995, when the Southern Baptist Convention (SBC) voted to repent and ask for forgiveness for their historical support of slavery and for racism within their ranks. With these high-profile events the spirit of the racial reconciliation movement became institutionalized.

And so white Christians in America entered into the twenty-first century with a new vision of themselves. Frances Fitzgerald (2017) has written about Rick Warren as an example of what she calls the "new evangelicals"—a movement that grew out of the disappointment and disillusionment evangelicals felt as they were increasingly identified as "the Republican party at prayer" during George W. Bush's tenure. Evangelicals realized they had lost the culture wars on many fronts, and these new evangelicals tried to reenergize evangelical culture by focusing on issues like poverty alleviation and climate change. It was in this spirit of rebranding that Russell Moore, the head of the SBC during this time, made "racial reconciliation" a central part of his platform.

But Moore's brand of racial reconciliation was shaken first by the rise of Sarah Palin and the "teavangelicals" (see Butler, 2021, pp. 114–125) and was finally laid bare in 2012. If Moore was the angel on the shoulder of the SBC, on the other side was Richard Land, head of the Ethics and Religious Liberty Commission (ERLC) in the SBC. Land also hosted a popular radio show. After seventeen-year-old Trayvon Martin was killed by George Zimmerman as the young boy walked home from buying Skittles, the country saw one of the first fronts in the contemporary fight for racial justice open up. On his radio show Land implied that Zimmerman was right to be afraid of the child, because, he claimed without evidence, Black men were statistically more violent than white men. Land also accused then president Barack Obama along with Al Sharpton and Jesse Jackson of being "race mongers" and using the child's murder to get Black votes. *Christianity Today* called Land's comments "a PR headache" (Feddes, 2012), and he was subsequently fired from ERLC. But even as the SBC attempted to paper over it, this moment revealed a deep division within the largest American evangelical institution over issues of racial justice.[12]

This division grew deeper during the presidential election season of 2016. Sarah Posner (2020) has made the case that Russell Moore's progressive stance on racial reconciliation put him at odds with the prevailing winds of white evangelical culture, winds that were shifting rightward, toward support of a Trump presidency. And John Fea (2018) has identified white grievance as a central driver of white evangelical political loyalty to Trump. Trump spoke directly to the concerns of white evangelicals and energized a base of Christians who, like Richard Land, remained insistent of the idea that Black people were overstating the racism inherent in American society and culture. Ultimately Russell Moore was sidelined when Trump ascended to power and Moore's vision for racial reconciliation was set aside as overly conciliatory. Trump rewarded his evangelical fans by promoting and spotlighting a cadre of "court evangelicals"—famous Christian media figures such as televangelical celebrities Paula White and Robert Jeffress.[13]

The increasing polarization of politics along racial lines made it difficult for the church to explain away racism or to see it as a problem that had been overcome in the 1990s. And yet white evangelicals seemed to double down on their insistence on colorblindness and their urgent call to Black Christians to stop making race an issue. Christian writer Austin Channing Brown has explained that white evangelicals "have allowed *reconciliation* to become synonymous with *contentedly hanging out together*" (Brown, 2020, p. 167, emphasis in the original). White evangelical institutions are happy to accept Black congregants—indeed they often seek them out. They hire Black pastors, musicians, administrators, and in doing so they hope to prove that they are colorblind and accept people from all races. But as the experiences of Brown and many other Black Christians show, this is a smoke screen behind which is a disciplinary culture that punishes Black Christians for expressing views counter to the political norms of white Christian culture.

As many Black Christians have explained, this attitude is often read as an insistence on assimilation. Ally Henny, host of the podcast *Combing the Roots*, made this point strongly in my interview with her: "Whenever we come into the white churches, we're expected to surrender our culture at the door. If not our skin color, definitely our culture. We're expected to assimilate and expected to do all these other types of things" (Henny, 2020). As Henny points out, in these spaces assimilation often means silence—a tacit agreement that Black Christians should not discuss causes of racial justice within white evangelical spaces because it will make white people feel uncomfortable. This attitude has led to what the *New York Times* called a "quiet exodus" of Black worshippers from white evangelicalism.[14]

As I show in my analysis, for Black Christians in a post-Trump world the discourse within Christian culture can no longer be centered on a 1990s understanding of racial reconciliation, nor can the emphasis be on creating multicultural churches. Black Christians say that the focus has to shift to racial justice and racial reckoning. They want white evangelical institutions to face the past and present racism in their ranks and to repent and repair for their historical support of white supremacy. And this chorus of Black Christian voices who had once been on the margins of the discussion are using the technology and media at their disposal to express their perspectives, share their stories, and create counterpublics.

But it would be wrong to assume that Black Christians are hopeful that white evangelicals are listening. Henny tells me that she does not think white evangelicals will change. "I just see it as an inherently bankrupt system, because it's not interested in dismantling white supremacy . . . [it] is interested in slapping a saccharine Jesus coating on an issue. The evangelical church, to be fair, does that with a lot of things" (Henny, 2020). For Henny, evangelical culture is too tepid, too fearful of upsetting white parishioners to meaningfully address race. All that it can offer, in her opinion, is lip service, "evangelical gentility" and not a commitment to racial justice or to disentangling white American Christianity from white supremacy. To achieve this, Henny and others are careful to explain, would take work. But, if white Christians wanted to change their hearts, their churches, their culture, Henny and other Black Christians tell me, they could start tomorrow. But, because it has failed again and again, the individualist focus of racial reconciliation that comes from white evangelical spaces is no longer enough, and Black Christians can no longer accept it. So what comes next? In the summer of 2020, during a crucial moment in American history, as protest marches churned and soldiers occupied American cities, Black Christians outlined a path forward for the American church, a path toward racial reckoning and repair.

PODCASTS AND THE BLACK CHRISTIAN COUNTERPUBLIC

Podcasts are prerecorded audio programs that are available to download on online platforms. While their format often resembles radio, many have argued that podcasting should be considered an art form in itself, currently experiencing a "golden age" (see Spinelli & Dann, 2019, p. 45). There are hundreds of thousands of podcasts of all types available to stream on services such as Spotify and Apple Music. Some podcasts have become immediate hits even been made into television shows (e.g., *Pod Save America, Homecoming, Lore, Sound Exploder*).

Both podcast listening and podcast production are aspects of digital habitus. Podcast listening is a habitus that implies that a person is trying to learn, educate themselves. There is an "epistephilia" at play here—to use Bill Nichols's (1991) understanding of the joy we get from watching documentaries, a joy in knowing. Podcast listeners seek out podcasts based on what they want to learn about and as such podcast listening practices reflect the tastes of individual listeners and the collective understandings of class-based listening practices. There's a podcast for every interest, and there are hundreds, if not thousands, of Christian podcasts. In some ways the most popular Christian podcasts reflect the most popular strains of American Christianity. Most of the faces one might conjure when thinking about popular evangelicalism (such as Joyce Meyer, Christine Caine, Rick Warren, Andy Stanley, and T. D. Jakes) have popular podcasts. But even those "outsider" evangelicals have used podcasting to grow their audience and communicate their ideas. Jen Hatmaker, for example, has a popular podcast called *For the Love* in which she discusses things that she loves and covers topics as diverse as "the enneagram" and "Black lives."

The rise of the *exvangelical* movement and its associated podcast provides an example for how podcasts have become sites for counterpublic dissent against the norms of evangelical culture. Blake Chastain first tweeted the term "exvangelical" in a hashtag campaign that highlighted the experiences of evangelicals who had left or become disillusioned with evangelical culture. Chastain turned this idea into a popular podcast that hosts conversations with people who have left evangelicalism. The tagline for the exvangelical podcast is "Coming to terms with a messed up subculture. One conversation at a time," and it reportedly boasts about thirteen thousand downloads per month.[15] Writing about the podcast and the associated movement, Bradley Onishi (2019)—who is himself an exvangelical and a popular podcaster—has explained: "Ex-evangelicals hold a singular potential for undermining evangelical politics. Those who associate with #exvangelical are not leftist outsiders with no real experience within the subculture. They are former insiders who testify to what they see as the traumatizing effects of living under evangelicalism's patriarchal, heteronormative, and racist norms." As the success of this podcast shows,

insider critiques of evangelical culture aimed at audiences who understand this culture, are part of it, or feel curious about it flourish in the
podcast scene. Like the hashtag movements chronicled in chapter 4, these
podcasts have the potential to connect and galvanize counterpublics by
highlighting and amplifying voices that speak out against the evangelical
power structure represented by Bible colleges, churches, and parachurch
organizations and other sites of cultural power.

Host Tyler Burns of the podcast *Pass the Mic,* which talks about
Christian culture from a Black perspective, tells me: "What I think podcasts have done in this kind of new way, especially in a post BLM [Black
Lives Matter] movement is just level the playing field between the status quo and the power structure within the church and given marginalized people within the church the opportunity to have a voice, but also
the opportunity to craft a new narrative" (Burns, 2020). Though popular
televangelists and megachurch pastors have a built-in audience for podcasts, newcomers can attract followers, create new audiences, and circulate ideas that come from the grassroots, rather than from the top of the
power structure. Burns is a pastor in addition to being a podcaster, and
he explains how podcasting can change the dynamics of the Christian cultural conversation:

> If you have a megachurch, now you, you're in a seat of power. If you're Rick
> Warren, or if you're a Joel Osteen, or if you're T. D. Jakes, Robert Jeffress or
> whoever you have the power and control, right? Because you control that
> narrative. Now podcasts have evened the playing field because we have just
> as many listeners as they do in the seats. And so now people are engaging
> a different narrative. They're not going to the pulpit first to hear shaping of
> the narrative. The pulpit is just one of the voices now. (Burns, 2020)

With widespread digital habitus, smartphone adoption, and platforms
like GarageBand and Adobe Audition, the barriers to entry for this broadcast medium are far lower than radio's ever were. If you have an internet
hookup and know how to look up YouTube tutorials, you can start a podcast. The people who produce podcasts display their fitness in a culture
that values digital literacy. Like Christian influencers on social media, they
are often performing their authenticity—that is, the more that listeners
feel like they know and can relate to the hosts and guests on the show, the

more they will want to listen.[16] But unlike the social media influencers posting photos of their "authentic" selves, this is a medium dominated by the voice and by discussion rather than by visual expression. It becomes clear when you listen to Christian podcasts that people trained as preachers are particularly suited to be podcasters as the rhetoric and logistics of a sermon adapt well to this medium where forty-five-minute to one-hour discussions focusing on a theme are the norm.

Podcasting as an extension of the broadcast medium of radio also affords it distinct advantages to Black podcasters. Namely, because it is a primarily asynchronous medium in which podcasts are recorded in advance and uploaded to distribution platforms, they cannot be disrupted in the same way that a social media conversation or even streaming video can. Thus podcasts are somewhat inoculated from trolling, whereas social media conversations and live videos might be easily derailed by trolls, podcasting is somewhat immune to trolls and this is important for Black Christians talking about race, because trolls often target online discussions of racial justice.[17]

So Black Christians can express ideas and share experiences in and through a medium that is dominated by white interests and white audiences without having to engage with those who might try to derail the conversation. One young podcaster, Michelle Lenae, tells me that what led her to create her *Christ over Culture* podcast in 2016 was her interest in hearing from the experiences of other young Black women. She recounts that she went to Apple Podcasts and "just saw a lot of white, Caucasian, either women or men, and particularly of the older generation, like Baby Boomers, giving that type of advice. And while there is a lot of wisdom in that group, I really wanted to create a space for a diverse community to come together and talk about some of these things that young adults, especially Black women go through" (Lenae, 2020). Lenae wants to carve out a space where people like her, young millennial Black Christians, can discuss the issues most important to them. And the first topic she covered was the Black Lives Matter movement from a Christian perspective, because, as she tells me, "sometimes in faith communities, racism isn't talked about, or is tiptoed around" (Lenae, 2020).

As they express the concerns, ideas, and discourses of their community, these podcasters create counterpublics that are similar to the counterpublics that cohere around female social media influencers but are also distinct from them. Catherine Squires (2002), who has written about Black counterpublics, schematizes these spaces in three ways: enclaved publics, counterpublics, and satellite publics. Enclaved publics remain shielded or hidden, perhaps, for example, to protect the safety of their members under conditions of repression. Satellite publics may engage with the broader public sphere, while remaining separate from it. By contrast, Black counterpublic spaces actively engage the broader public sphere as a way to force change within it. Sarah Florini's (2015) study of Black podcasters reveals the power of the medium as a site of enclaved discourse. As she argues, the style and personal interaction that the medium mimics resemble the experience of in-person Black sociality. On the habitus of podcast listening, Florini surmises that "many of these listeners consume podcasts via headphones in predominantly white spaces where they work or live, and may to some degree be 'cocooning' themselves in the sounds of Black sociality as they navigate a hegemony that constitutes white culture as normative" (2015).

André Brock (2020) has theorized that "Black information technology use highlights Black technical and cultural capital while disrupting the white, male, middle class norms of Western techno culture. Black digital practice challenges these norms through displacement, performativity, pathos, and the explicit use of Black cultural commonplaces" (p. 17). Brock argues that Black users have carved out Black spaces in a digital culture where the assumed user is a white male and they have done so by performing Blackness. In this process, networked Black users have defined Black technocultures, which, Brock writes, connect Black cultural practices to information technology use, or to use my term, digital habitus.

Although Black listeners can retreat into Black spaces through the habitus of podcast listening, as Florini and Brock argue, the podcasters that I spoke with for this chapter also share the goal of changing Christian culture, which means talking to and trying to influence white Christians as well. Black Christian podcasters are creating both an enclaved space where Black listeners can feel at home and cocoon themselves in Black cultural idioms, and a counterpublic space that engages with the larger Christian public. And this dual nature of the medium makes it powerful. As scholars of digital media have theorized, individuals understand that when they communicate through social media or other new media technologies that they may be speaking to multiple audiences. This has become known as "context collapse," and empirical studies show that users internalize this understanding of multiple audiences and it influences the manner of their speech and what they speak about.[18]

The public sphere of evangelicalism, which is comprised of publishing houses, magazines, television shows and channels, and podcasts, has been controlled by the mostly white evangelical gatekeepers of the Christian media world. And many of these gatekeepers have proven that they are willing to block voices that do not conform to what is considered the norms of white evangelicalism.[19] But with the rise of podcasts, Christians can sidestep these cultural authorities; they can create spaces that are defined in and through the Black experience and read to Black Christians as safe, enclaved publics, but they can also use these conversations as a space to push their own discourses into the broader public's consciousness, which makes them counterpublics.

The podcast *Pass the Mic* had a primarily white audience at its inception because it became popular among white Reformed Christians after it was featured at the 2013 national conference of the Gospel Coalition—a highly influential event in modern evangelicalism that typically boasts around ten thousand registrants.[20] But Tyler Burns tells me his intent was not to speak to a large, diverse audience. In fact, the goal of the podcast was to create a space for Black Christians—an enclaved space—and he was intentional about performing in a way that would make Black audience members feel comfortable.

> I want to talk to Black Christians. I want to speak to Black people. . . . I want
> to code myself as Black because here's what happens if I speak to Black
> Christians directly, there are other Christians that can feel as though, okay,
> I identify what's being said, but if I speak to white Christians directly, then
> Black Christians will feel excluded and alienated and further marginalized.
> (Burns, 2020)

But even as they work to carve out Black spaces in digital culture, these podcasters also share the desire to change the broader Christian public. These podcasts are public facing, and their hosts and producers understand that white evangelicals are listening. Jemar Tisby, Burns's *Pass the*

Mic cohost, explains that it is crucial that *Pass the Mic* is a space made by and for Black Christians, but "sometimes we speak directly to white evangelicals because there's a lot of them they're oftentimes part of what impedes Black freedom in the church or beyond, so in my view, they need to be in the conversation. There needs to be conversations that are tailored to their misunderstandings and their proclivities" (Tisby, 2020). Here, Tisby describes his rationale for addressing white evangelicals. He knows that true change in Christian culture needs to include them if only because they represent such a large swath of Christians in America and because they control many of the institutions—Bible colleges, churches, parachurch organizations—that define Christian culture in America.

Below I sketch out the alternate vision for racial healing constructed by the discourses that Black Christian podcasters broadcast and engage in. This is a vision for racial reckoning, as opposed to racial reconciliation, but at its most hopeful it imagines a united Christian church in America fighting against and overcoming systemic racism in the church and creating an interracial coalition of believers that could provide a moral vision for the country. The discourse circulating in the Black Christian podcast scene is constructed by Black Christians, it is an enclaved space tailored to their interests and experiences, but it is also a counterpublic that hopes to influence white evangelicals and offer them a path forward—if they are willing to listen.

DECOLONIZING FAITH

"Is Christianity a white man's religion?" Lisa Fields asks her guests on the *Jude 3 Project* podcast. The tag line for her podcast is "helping you know what you believe and why," and it focuses on Christian apologetics that seek to correct what Fields sees as common misconceptions about Christianity in the Black community. Many African American view Christianity as the religion of slavery and of colonial oppression and understand their participation in it as participation in an oppressive historical lineage. One of the podcast's recurring themes is the idea of decolonizing the Christian faith for Black Christians. As Fields explains it, she wants to offer a blueprint for Black Christians to understand a precolonial expression of Christianity with its roots in the African continent. She does this by interviewing scholars and theologians whose work concentrates on African Christianity. As Black podcasters work to create enclaved spaces for Black audiences, addressing the concerns of these audiences is key, and a focus on decolonization is a means to that end.

Franz Fanon (1963/2004) wrote about the process of decolonization, explaining that "what is singularly important is that it starts from the very first day with the basic claims of the colonized. In actual fact, proof of success lies in a social fabric that has been changed inside and out" (p. 1). In the U.S. context the term has come to signify participation in counterdiscourses that seek to reclaim Black experience and Black identity. Decolonization in this sense implies a project of reclamation in which Black Christians might recover some of what is lost in the long historical, often physically and psychically violent processes of cultural assimilation.

But the question "Is Christianity a white man's religion?" is also indicative of the implicit bias of white evangelical spaces who see American Christianity as the norm and who might discipline or deride those whose expressive or liturgical norms are apart from this. In asking this question to theologians, Christian celebrities, and others, Fields challenges her guests to reclaim historical and theological narratives and recenter Black theology and Black liturgical expression in Christianity. About the process of writing his influential book *Black Theology and Black Power* (Cone, 1969/2018) directly after the assassination of Martin Luther King Jr., James Cone noted: "I had to deconstruct white theologies to destroy their effects on my mind so that I would be opened to listen to the Black voices from slavery, emerging from the ashes of the Black holocaust. I had to look back and *recover* the Black heritage that gave birth to me" (emphasis in the original). Here we see Cone describing how he actively practiced decolonization in his own life and writing. In other words, we see him writing his way out of the DuBoisian "double consciousness" characteristic of the Black experience in America.

W.E.B. DuBois first wrote about the idea of "double consciousness" in 1897, just three decades after the Civil War. He defined the experience of being Black in America at that time with regard to the idea of "two-ness": "this sense of always looking at one's self through the eyes of others, of measuring one's soul by the tape of a world that looks on in amused contempt and pity. One feels his two-ness—an American, a Negro; two souls, two thoughts, two unreconciled strivings; two warring ideals in one dark body, whose dogged strength alone keeps it from being torn asunder" (DuBois, 1897, p. 22). More than a hundred years later, Black Christians still express this sense of double-consciousness, especially those who worship in white evangelical spaces, because white evangelicals represent a particularly conservative segment of American society and Black Christians recognize this and feel surveilled by it.

Because they are embroiled in this situation that makes them feel as though they have to be careful about what they say, how they dress, and how they express themselves so that white Christians will not fell threatened by their presence, Black Christians often note that they do not feel free in white evangelical spaces. Ally Henny (2020) recalls that when she started talking about and participating in racial justice activism: "I knew instantly that, that put me at odds with every person that I had attended church with in the past decade." In her experience the white evangelical church was silent about race, but they also tacitly required that Black Christians too remain silent about race. As soon as Henny started posting on social media about Black Lives Matter, she says she received condemnation and threats from people she thought were her friends. Black Christians understand evangelical churches as raced spaces. These churches represent a Christianity that has ties to white Christian nationalism and sometimes that rhetoric is explicit and violently deployed. As many Black Christians have explained, this has become clearer and clearer through the Trump presidency and the Black Lives Matter movement.

But even for those Christians who worship in Black churches, there remains a concern that the traditions of Black Christianity reflect a colonized mind-set. In seminaries and in white evangelical culture, the traditions of the Black church are often characterized as other or lesser. Dr. Vince Bantu, who specializes in Black church studies, was a guest on the *Jude 3 Project* podcast and expressed the idea of "good negroes" and "bad negroes" in early Christianity. Those African theologians who wrote in Roman and Greek were embraced by the Roman Church, but those who did not were violently suppressed. And this ancient schism has ramifications for Black Christians in America today. Bantu connects the way that European theologians spoke about African theologians to the schism between white and Black churches today. He asks the podcast audience: "How many of us Black Christians today have been told that Black preaching is less theologically robust and less theologically profound than white preaching or that gospel music is low worship?" (as quoted in Fields, 2020a).

Bantu is referring to an experience that came up in other podcasts and in my interviews with Black Christian media makers. For example, many ordained ministers recalled how they were trained in white institutions where the liturgical traditions of the Black church were characterized as less theologically sound than those of white evangelical culture. Bantu connects this to his own coming-of-age story. He says he was taught to see the Black culture of his youth as evil: "I thought to follow Jesus you had to fully reject who you are" (Bantu as quoted in Fields, 2020a). By connecting ancient Christianity to the racial politics of the United States today, Bantu and Fields are doing the work of decolonizing Christianity for Black Christians by giving them a historical lens that does not rely on white or European interlocutors. They tell Black Christians that they do not have to see themselves through the lens of white theologians or a white evangelicalism that refuses to accept their culture and liturgy.

"Decolonizing Christianity" means blazing a trail for Black Christians to follow that takes them on a path toward recognizing a Christianity nondefined by whiteness or Americanness. And in so doing, it creates an enclaved space for Black Christians to celebrate Black Christian expression. This sets the stage for all of the discourses to follow. This discourse needs to be led by Black Christians, these podcasts assert, and they need to express a Black Christian worldview that is defined by Black expression and a Black history that reaches back into the African continent. Lisa Fields hopes to take Black Christians to this place, a place from which they might speak in a united voice about the problems in American society and in Christian culture. Before she can get there, however, she has to make the case that Christianity can be seen in an intersectional, international, and diverse way.

RACIAL TRAUMA

Across all of the podcasts I listened to, I heard Black Christians express and explain formative experiences of racial trauma in their lives. Some recalled the feeling of being one of a handful of Black students at their elementary school, their high school, their Christian college, or their seminary, and they spoke about microaggressions and acts of discrimination and racism perpetrated against them by individuals and by the oppressive norms of white institutions. They spoke about how they were taught, for example, that hip-hop was demonic, that Black cultural expression was wrong, and they modeled their behavior on these assumptions, sometimes accepting their own repression for a time in order to fit in to Christian spaces.

Black Christian podcasters often highlight the importance of therapy, which is something that is not always accepted in Christian culture. Prominent podcaster Dr. Thema Bryant-Davis, who is both a psychologist and an ordained minister, tells me: "Many people in the faith world have been discouraged from going to therapy or even acknowledging mental health problems. And so this idea of like, if you love God, you'll always be happy. Right. You love God, you'll never have anxiety and it's not even biblical" (Bryant-Davis, 2020). Because it is not widely accepted in Christian culture, Black podcasters' continued insistence on therapy stands out as a way that this counterpublic specifically acknowledges the racial trauma that Black Christians experience in white spaces.

Scholars, journalists, and pollsters often refer to white evangelical churches as though these are truly monoracial spaces. That's not the fully story. White evangelical spaces remain largely white, Black Christians will tell you, because Black Christians are often subjected to racism within those institutions. In other words, they are forced to leave, or they are forced to stay silent if they want to stay. But Black Christians cannot completely escape white evangelical culture. Most Bible colleges are run by white Christians; Christian nonprofits in which Christians of color might seek employment are similarly dominated by a white evangelical ethos.

Cohosts Burns and Tisby of *Pass the Mic* reference their respective histories in white evangelical institutions and the trauma that these institutions caused to them, and they directly address Black Christians who may be part of white institutions. For example, in one episode Burns forcefully states: "Black Christians we see you. Get out of spaces that are oppressing you" (Tisby & Burns, 2020b). And they clearly lay out the stakes for Black Christians. In another episode Burns offers a blueprint for how they might protect themselves in white churches:

You need to go to your church. You need to write an email. You need to ask what is the policy for dealing with racists in this body. If you are in a multiethnic church there needs to be a policy. If someone responds any racism towards you in person, via a text, via messenger, or on social media what is the policy? Follow up question: if this happens to me, if this happens to my child in student ministry, if this happens to my child in kids' ministry, if this happens to my wife, if this happens to my husband, is the burden of confrontation going to fall on my shoulders? Am I responsible for confronting this person? Because here's the thing—this isn't some sort of random dispute that we are going to work out. No, this is trauma to my body. It's emotional trauma. I need you to understand the power dynamics. I'm at a power disadvantage. (Tisby & Burns, 2020d)

Burns goes on to say that churches need to have Black mediators from outside of the community who can offer protection for Black congregants who may face racist aggression in white spaces. He implies that this is somewhat inevitable—Black Christians should anticipate racial trauma will be inflicted on them. And so, he says, they need to think ahead, about the potential consequences of trauma by asking church leaders: "Is there a fund available so that me and my family considering that there may possibly be a situation where we will be harmed and traumatized racially—is there a fund for racial trauma counseling? Is there a fund for therapy?" (Tisby & Burns, 2020d). Burns lays out these questions forcefully. They come from an understanding that white evangelical spaces have not taken racial trauma seriously, and because of this, they continue to inflict it on Black parishioners.

Mutual forgiveness has been a central pillar of the evangelical approach to racial reconciliation. And it carries with it a particular visual and emotional appeal. A white pastor can give a sermon on racial reconciliation in which he performatively washes the feet of a Black parishioner in the manner of Jesus washing the feet of his disciples in the Bible. This image provides a powerful semiotic stand-in for racial unity. But, as Christian counselor Sheila Wise Rowe (2020) has written about racial trauma in the United States, many people have grown weary of this type performative repentance:

People of color have received apologies, but true reconciliation involves the perpetrator's personal and public apology to people of color for the past and present harm. Also, *repentance* must include a commitment to stop all actions of injustice and oppression and to repair any damage done. If we want true reconciliation that is flourishing, then *repair* is essential; without it the apology and repentance can feel shallow. (Wise Rowe, 2020, p. 137, emphases in the original)

Despite attempts to visually or semiotically distance themselves from the racism of the past, such as the on-stage embraces between Black and white Christians during the Promise Keepers' heyday, Black Christians do not believe that white evangelical spaces are doing enough. They say, forgiveness is not enough. Nor is it enough to hire Black staff, to have representatives from the Black community as visible signifiers of reconciliation in white evangelical churches. Instead of forgiveness based on a flawed understanding of racism as an individual sin, these Black Christians are asking for repentance, repair, and sometimes reparations. Tyler Burns talks about the emotional performances of repentance he finds all-too-common in white evangelical spaces: "Why are you crying? Write a check! Repair don't just repent" (Tisby & Burns, 2020b). For Wise Rowe and Burns, apologies ring hollow after decades of continued trauma. What they point to is the need for white evangelical spaces to go farther, to find ways to *repair* the trauma, such as providing financial support or counseling for Black parishioners or financially supporting Black ministries.

In addition to the trauma of erasure, nonacceptance, and racism in white spaces is the trauma of seeing images of Black people murdered or violated by police in the media. Allissa V. Richardson (2020) has argued for the concept of Black witnessing, by which she means the way that Black people view images of brutality perpetrated against Black people. The recurring trauma of watching media reports and viral videos of racist violence perpetrated against Black people creates a particular subjectivity and a particular mode of witnessing, according to Richardson. Where a white person might watch footage of police brutality against a Black person and feel indifferent, or feel like justice is being done, or feel like there may be things that the video is not showing that explain the police officer's behavior, a Black person does not see this footage within these frames. Instead, these videos spark a need for action in Black viewers. However, these images are also traumatic and damaging to the psyches of Black people.

"Bearing witness while black, after all," Richardson writes, "is gazing into forbidden space—the space of vigilante and state-sponsored violence against black bodies. It is a ferocious space that many African Americans always knew existed" (2020, p. 203). While facing this reality galvanizes people to act, to create counternarratives, to protest, it also creates a baggage that many Black Americans carry. But many Black Christians have found that they cannot carry that burden into white evangelical spaces. Black Christians talk about being "gaslit" (being made to feel crazy) in white evangelical spaces in which they are told that they are overreacting to images of police brutality against Black citizens on the news, or that the images do not represent a real problem that needs to be addressed.

This disconnect between Black and white witnessing of mediated, racialized violence creates a cultural chasm in the church, one that white Christians, on the whole, have failed to acknowledge. Ultimately this aspect of racial trauma is doubly damaging for Black Christians in white evangelical spaces because the churches and institutions that inflict this trauma often do not acknowledge its validity. Over and over again I heard Black Christians explain that their final breaking point, the moment that they realized they had to leave their primarily white church or cut ties with white evangelical friends was one of three events: the murder or Trayvon Martin (2012), the killing of Mike Brown and the subsequent protests in Ferguson, Missouri (2014–2015), and the election of Donald Trump (2016). In these few short years, at each of these points, Black Christians saw white Christians continually take stands against racial equality en masse. Black Christians saw prominent white Christians denigrate Black victims. This happened in churches, in seminaries, in Christian institutions, and in Christian media. And what this revealed, as many Black Christians have explained, is that white Christian America was (and is) unwilling to see Black people as fully human and worthy of empathy.

Tyler Burns clarifies the situation many Black Christians find themselves in: "They are killing us in the streets and we having to watch it and then go explain it to people who aren't willing to do the work" (Tisby & Burns, 2020d). When Burns talks about white Christians who are not willing "to do the work," he expresses a sentiment that I often heard in my interviews and analysis: white evangelicals are not willing to educate themselves about racism. In fact, many white evangelicals believe that racism is not the problem that Black Christians believe it to be, and they will publicly (on social media and other platforms) deride the Black Lives

Matter movement without considering how their disdain for racial justice movements affects the Black Christians they may be friends with.

One high-profile example of this pattern in white evangelical culture to accept Black people but to discipline them when they support racial justice activism comes from the Christian hip-hop artist Lecrae. Lecrae began making music in 2004 and as his career grew, he became a popular performer on the Christian conference circuit. But as much as he was sought after in white evangelical spaces to perform, Lecrae was criticized by white evangelicals for his support of Black Lives Matter. In an open letter from 2016 he explained:

> I hit a serious low on tour at one point; I was done with American Christian culture. No voice of my own. No authenticity. I was a puppet. I'd seen so much fakeness from those who claimed to be my brothers and sisters that I didn't even know how to talk to my Heavenly Father.
>
> And then there was Mike Brown.
> Then Eric Garner.
> And then #_____ and #_____ and #_____.
>
> People kept killing us.
>
> As I shared my heart, my supporters turned on me even more—fans and friends. There was no empathy. Though some comments were just evil and hurtful, others were steeped in ignorance and lack of perspective.
>
> They didn't get it. (Lecrae, 2016)

Lecrae's experience is one echoed by many Black Christians worshipping or working in white Christian spaces. They see themselves as doing the important work of representation, they are often hired because they are Black and because white Christians want to diversify their congregations. But when they express support for racial justice, they are criticized and sometimes fired.

On the *Pass the Mic* podcast the hosts regularly highlight stories from Black Christians who have left white institutions, sometimes devoting entire podcast episodes to #LeaveLoud, a hashtag campaign meant to highlight Black people leaving white spaces in a public way. One particularly poignant story featured on *Pass the Mic* was that of a Black woman named Zee who served as a musician in a primarily white church in Texas.

Zee was fired on a Saturday night before she was supposed to perform on Sunday because of a Facebook post in which she had expressed the need for pastors to honor racial justice initiatives. As Zee put it, she was not only fired from her job at the church, but she was excommunicated from her church community. She felt as though "my heart had no home" (as quoted in Tisby & Burns 2020c), and she described the subsequent emotional trauma she experienced as she suddenly found herself in exile.

I heard similar stories from interviewees and recognized the effects of this trauma in the emotion in the voice of a women at a conference who asked a speaker how she might forgive the people in her former church, white Christians whose racial insensitivity caused this woman to leave her congregation. White churches want their churches to look diverse, but Black Christians feel that church leaders have proven unwilling to accept ideas or behaviors that they feel will alienate their largely white populations. They still express the "evangelical gentility" of Billy Graham that allowed for white and Black Christians to sit, sing, and worship together—as long as no one upset the white status quo. When given the choice between doing the work of educating themselves and their congregations on racial justice or disciplining Black congregants who speak about it, Black Christians have seen white church leaders choose the latter all too often.

THE NEW CIVIL RIGHTS MOVEMENT AND CIVIC EXPRESSION

The Church Politics Podcast, helmed by Justin Giboney and Michael Wear,[21] is a wonky, policy-obsessed podcast that attempts to model a mode of Christian political engagement that might transcend party politics. And in its conversations the podcast urges Christian people, organizations, and churches to become more engaged, especially on issues of social justice. *The Church Politics Podcast* clearly lays out ways that Christians can disentangle themselves from the toxic nature of party politics in favor of discovering a more authentic Christian political lens.

Though the podcast is invested in disentangling the polarized political discourses in which Christians often find themselves in favor of offering a third way, on June 10, 2020, in the midst of the uprisings against police brutality taking place in cities across the United States, Giboney was clear about the role that Christians should play in the right for racial justice:

> If the whole church would have stepped up years ago we wouldn't be in this dire situation. Christians who supported slavery, and then supported Jim Crow, and then supported redlining—if the Christians who just went along with that would have stepped up years ago, we would not be in this position. We cannot run away from that history, we have to be real about that history. Christians have failed in many ways to do what we needed to do and we would not be in this place if we had been the people we were called to be. (Giboney & Wear, 2020)

As Giboney put it, the Christian church in America is a powerful cultural force, and it can be an agent of change. Yet too often in history Christians have used their power to maintain the often violent and unequal status quo. But even if the church has been complicit in the past, Giboney continued, there is an opportunity from Christians to start right now. And the way to start, he said, is through Christian unity on the issue of racial justice.

All of the podcasts I analyzed for this chapter expressed palpable urgency in the summer of 2020 when there were protests in every U.S. state after the deaths of George Floyd and Breonna Taylor at the hands of police, and Ahmaud Arbery at the hands of white supremacists. Podcasters spoke of the potential for change and the role that Christians in America might play in clarifying the moral vision of the moment. In their conversations they strategized, they lamented, they organized, and they thought back to the moments and ideas seared in the collective memory Black Christians—the Civil Rights movement and especially the role that the Black church played as an organizing and galvanizing force in Civil Rights activism.

These podcasters all expressed the idea that the Black church needed to be more engaged as it was in the days of Civil Rights, when protests were organized in churches and famous preachers like Martin Luther King Jr. communicated a Christian vision of racial equity. A speaker at a conference I attended put on by the *Jude 3 Project* remarked: "One of the saddest thing about the Black Lies Matter movement is that it did not start in the church." During the height of the uprisings, Chicago-area pastor Reverend James Meeks mused on another episode of the *Jude 3 Project* podcast: "Could you imagine if every church rather than singing ninety-four praise and worship songs on Sunday, if we just sang ninety-three and took five minutes say, Okay, we're going to talk about the police contract?" (as quoted in Fields 2020b). Here Reverend Meeks expressed one way that the Black church might step up and get more involved in the Black Lives Matter movement.

But even as they criticize the Black church for not doing enough to live up to the memory of the Civil Rights movement, these podcasters also express anger that the white evangelical church seems to be doing less than nothing. During the height of the uprisings in the summer of 2020, for instance, there were calls for unity among Christians on the Black Christian–focused podcasts that I analyzed. But this was a markedly different vision of unity than that offered by the Promise Keepers. In 2020, Black podcasters hoped to see a united church express a moral vision for racial justice from an interracial coalition of believers. The podcasters whom I spoke with for this chapter shared the goal of unity in the church. They want to change the conversation happening in mainstream evangelical culture because they believe that the church, when united and working together, can be a powerful change agent. In my interview with him, Tyler Burns (2020) put it this way:

> What I desire to see is for repentance and repair to start in the church, repent and repair within the church first and that the model and example of repentance and, and reparation within the church, equips and mobilizes a generation of Christians to pursue the same thing in society. And so the reason why we're doing this and pushing for this in the church is number one. Because we love the church and care about the church and believe that the only way we'll see the true beauty of the church as Christ has intended it to be seen is if we repent and then repair, if we correct the wrong that has been done to acknowledge it and then work to correct it.

The vision for racial reconciliation that Burns outlines sees Black and white Christians modeling productive, peaceful race relations for the rest of the country. This vision recurred in my research often self-reflexively expressed as a lofty, most-likely unattainable hope but also as a central goal of this counterpublic. The church should be a model for what racial reconciliation can look like, Black Christians say, but if it is, it cannot look like the racial reconciliation movements of the past that were led by white Christians.

First, Black Christians say, there needs to be an acknowledgment of the role that the church played in the institution of slavery. The collective memory of Black Christians is radically different from that of white Christians. And because of this, white evangelicals need to be willing to learn about Black history and the Black experience in America. White Christians need to meaningfully and publicly repent for the church's role in upholding the institutions of slavery and white supremacy. Next, there needs to be a repudiation of Donald Trump's inflammatory rhetoric. Black Christians were clear on this point in my interviews and across all of the podcasts I analyzed. They saw white evangelicals parroting Trump's divisive rhetoric on matters of race and understood that there could be no unity in the church if white Christians accepted Trump's version of history and of the current American experience—a vision that stubbornly refuses to acknowledge any implication that systemic racism exists in the United States. Finally, there needs to be unity on specific policy issues that Christians can provide a clarion moral voice on. Issues of police and prison reform, Black Christians voice, should be central to a Christian ethic. Black Christians imagine a united church whose activist energy could be put to use fighting for reform in these areas much in the same way that Christians, on the whole, have united in their antiabortion activism.

Black Christians understand that the future for racial justice has to be led by a diverse coalition of Christians—they believe in the power of the church. And their podcast conversations urge the church—conceived of as a united body of believers—to act. In this way these enclaved spaces speak as a counterpublic aimed at a the larger Christian public. But Black Christians have grown frustrated with white Christians, who seem to be working against them. Even those who are allies are not doing enough. In one two-part episode of the *Jude 3 Project*, Lisa Fields brought together panels of scholars and activists to discuss how to move forward. On this episode Michael Wear, of *The Church Politics Podcast*, noted: "We don't need any more allies, we need accomplices" (as quoted in Fields, 2020b). Black and white Christians need to be actively working together to fight for racial justice, Wear explained, rather than simply standing hand in hand. On the whole, he conceded, Christian accomplices have not come.

Black Christians understand that the leadership of the Black Lives Matter movement is not coming from the church; instead, it is coming from the younger generation and from trans people and gay women. And the Black Christian podcasts I listened to expressed that an awareness of intersectionality[22]—the idea that individuals and social groups face oppression on multiple fronts and these oppressive matrices shape their subjectivity—was central to forming a multiracial coalition of Christian believers. Although the podcasters I spoke with and listened to are excited about the possibility of breaking down the sexist barriers of the church, they also understand that there is a long way to go.

On March 25, 2017, just a month after airing their first podcast, the three women who comprise *Truth's Table* hosted an episode that centered around "gender apartheid" in Christian culture, at Christian conferences, and in the Black church. At one point host Michelle Higgins brought up the idea that "ordainable" in the context of most Christian denominations simply means possessing a penis, yet this idea excludes women from participating in church power structures. Higgins joked that she needed a "penis-shaped microphone" to be able to speak in church (Higgins, Edmondson, & Uwan, 2017). On social media, women erupted with praise for the way that this episode called out sexism in Christian culture. Social media users called it "healing" and tweeted: "Thank you. Thank you. Thank you." On April 17, 2017, the Twitter account associated with the podcast tweeted: "We know the Gender Apartheid is our listeners favorite episode."[23]

But one white Presbyterian pastor, Todd Pruitt, responded on his blog by calling on his readers to report the podcast and the women of *Truth's Table* to their ecclesiastical authorities for punishment. He wrote: "I would encourage you to make the pastor and session of your church aware of this and encourage them to take action by contacting the appropriate churches and presbyteries. . . . Let us do all we can to keep these unbiblical ideas from spreading any further in the PCA and corrupting our Lord's church."[24] Pruitt posted links to various denominations the women of *Truth's Table* are affiliated with. Other pastors publicly objected to the penis joke or to the perceived endorsement of liberation theology—the theological stance that seeks to engage with and combat social and economic injustices as part of the Christian mission. Emma Green (2017) wrote about this skirmish in *The Atlantic*, noting: "The way it unfolded—online, across multiple platforms and cities, with pastors unafraid to unleash their harshest words from behind glowing screens—wasn't random, either. For many religious communities, this is what theological debate looks like these days." As with Jen Hatmaker, the knee-jerk reaction from many conservative white Christians was to take to their blogs and to appeal to the authorities to keep these women in line.

Patricia Hill Collins (2000) has written about the historical suppression of Black feminist thought and how it has been circumscribed within nesting systems of oppression from the larger political and economic social structure to such institutions as the Black church: "Rather than seeing family, church, and Black civic organizations through a race-only lens of resisting racism, such institutions may be better understood as complex sites where dominant ideologies are simultaneously resisted and reproduced" (Hill Collins, 2000, p. 86). Here she explains what the women of *Truth's Table* expressed in the "Gender Apartheid" podcast episode: churches are patriarchal spaces where women's voices and women's knowledge are less valued than that of men.

As discussed in chapter 4 in relation to evangelical culture, women in the Black church tradition have also used the spaces that they were allowed to occupy to exercise their own authority in church spaces, but today the racial reckoning that is being outlined by Black Christians has become more explicit in its call to break down the gender barriers in American Christianity. As with Christian influencers, these podcasters use the media to teach and preach outside of the church walls because they are rarely allowed to preach within them. Dr. Thema Bryant-Davis (2020) told me that when she was pursuing her master's of divinity with the intent to preach in her home denomination where her mother was also a minister: "I had a classmate, a white male, and he said, 'Well, you'll probably still get to go to heaven, but your crown won't have as many jewels because of your disobedience.'" Although Bryant-Davis laughed off this experience as purely ridiculous—"I don't need bling because I'm in Heaven. I don't need a blinged out crown!"—it reveals the tricky place that Black women find themselves in when they enter the halls of evangelical power. Even if they come from a denomination that does allow women to preach in church, evangelical culture, as a whole, frowns on the practice.

Ally Henny had similar experiences at Fuller Theological Seminary, which was part of what led her to become a podcaster. In Christian culture, she notes, "Black women are "not given the opportunities to be in some of these spaces. And then we create our own spaces. And then we're told, 'Well, your spaces aren't good enough'" (Henny, 2020). The podcasting circuit is an outlet for women to discuss their experiences as Black women in Christian spaces and to speak and make connections with other women. But as Henny alludes, when Black women turn to media making, they face the charge that what they are doing is inadequate or unsanctioned.

Despite the pushback, however, Black Christian women like Ekemini Uwan, Lisa Fields, and Ally Henny are leading conversations on the podcast circuit. *Truth's Table's* audience has only grown since their "Gender Apartheid" episode caused such an uproar—their episodes as of 2021 have been downloaded two million times. Through the podcast medium Black Christian women have been able to amplify their voices and their message. They have emphasized the need for the contemporary fight for racial justice in the church to include a focus on intersectionality and on the importance of gender equality in the Black church. But there are still historical, institutional, and cultural barriers to this kind of sea change in thinking about sexism and gender discrimination that in some ways mirror the barriers against racial equality in the American Christian church.

DIFFERENT THIS TIME?

"This time feels different." That was the refrain from many social media observers during the wave of protests that followed the death of George Floyd. "This time feels different," people tweeted under videos of conservative, Mormon, politician Mitt Romney marching in a Black Lives Matter march organized by Christian groups in Washington, D.C. "This time feels different," people wrote in the comments under reports that the white megachurch pastor Joel Osteen had marched with Black Lives Matter protestors. Around the same time, the president of the SBC posted a video in which he said unequivocally "Black lives matter."

Things felt different this time.

In the midst of this, podcast host Jemar Tisby (2020) tweeted: "Maybe 'something different' this time isn't happening 'out there' but in us. The measure of change is not simply what happens externally but what happens internally. That is not all the change we seek but it is essential change." The sustained protests that followed George Floyd's death opened up spaces for Black Christians to talk about their experiences in the evangelical church. And it seemed as though there were flashes of hope that there might be a united front of Christians organized against racist violence.

But then there was Louie Giglio, a white megachurch pastor who went viral when he characterized slavery as a "white blessing." And in the height of the protests against racism in the United States, the popular white evangelical author and radio host Eric Metaxas used his platform to assert that Jesus was white. And there was Madison Cawthorn, who used the moment of his 2020 election to Congress to tweet "cry more libs" as an act of defiance and also perhaps as proof of his white evangelical bona fides. All along the way there were social media posts, live videos, YouTube reactions, and Instagram posts from ordinary people entering the conversation around racial justice in the church and trying to derail it. The morass of conflicting facts and opinions, mudslinging, and whataboutism served as a reminder that the political polarization of the United States still guides the racial rifts in church culture. In my interview with Tisby in the summer of 2020 he talked about the "whitelash" that was coming. Even as public opinion seemed to side with those protesting for racial justice, he knew that there would be a commensurate and possibly violent reaction from the white establishment. Others I spoke with during this time wondered if there would be a realignment of denominations, as schisms between those Christians who believed in the premises of the Black Lives Matter movement and those who did not grew in their impasse.

Something is different this time, but it remains to be seen whether some white evangelical spaces will double down on the problematic racial politics that they have preached for decades, while others make racial reckoning central. There may be new schisms already forming. But as the conversations that happened in the summer of June 2020 on the Black podcast circuit show, Black Christians have seized the moment to offer a critique of racism in the Christian church in America, and they have outlined a path forward for white and Black Christians alike. They speak about decolonizing the Christian faith by illuminating Black and African expressions of Christian theology and worship, and they have placed an emphasis on repairing racial trauma caused by white evangelical indifference and discipline in white Christian spaces. Black Christians talk about uniting the church to make it a moral beacon for the United States to follow. And they have also turned inward and offered a critique of their own gender norms and have amplified those doubly excluded in Christian culture: Black women.

Their path forward does not rely on and does not make concessions for the egos of white evangelicals; it is an unapologetically Black perspective, unlike the racial reconciliation movement that preceded it in the 1990s. Because the medium allows podcasters to speak with the dual purpose of creating an enclaved community and an influential counterpublic, this discourse has remained centered on the Black experience of Christian culture in America, and it expresses a vision that comes from that place, without reliance on white evangelical affirmation. This is also the central goal of The Witness Foundation, the brainchild of Jemar Tisby of *Pass the Mic*. This foundation has raised over $1 million to support Black Christians in ministry endeavors and to give them resources so that they do not have to rely on white Christian foundations, businesses, or gatekeepers. Podcasting as a medium has allowed for the same type of grassroots network-building and organizing. While it is difficult to measure how these conversations might translate into social action, the fact that they are happening is important because they reveal how the authority structures in Christian culture can be challenged using digital media technologies that empower the grassroots. As Tyler Burns (2020) puts it:

> The greatest concern of white supremacists, white Christian nationalism, is that Black people and Black Christians will control their own narrative. And that's when we become dangerous. And that's when we become free is now we can tell people, no, what they said is actually wrong. No, that's wrong. And then they can't shut us down. And that's why it's such a threat yet at the same time, such a powerful medium as well.

The revolution might not be televised, but it also can't be turned off.

Conclusion

QAnon may be America's newest religion. It's cosmology is based on an alternate reality in which a cabal of elite politicians and wealthy celebrities kidnap and violate children and perform Satanic rites, and the only person who can save these children is Donald Trump. "Q drops"—the way that the anonymous leader communicates with his followers—started on 4Chan, then migrated deeper into the net to a site called 8kun. These cryptic messages, videos, or images prompt readers to go on virtual scavenger hunts and collect more "clues" that ultimately lead to predictions and prophecies. Most of these prophecies—like the idea that John Kennedy Jr. is still alive and would announce his intention to run as Donald Trump's running mate in October 2020—have never come to pass, but true Q believers see this as proof that malevolent forces are thwarting the correct order of things.

QAnon may be America's newest religion, but there is also evidence that it is spreading in American evangelical Christianity. Some megachurch pastors have promoted or endorsed it, and while more mainstream evangelicals vocally condemn it, QAnon continues to attract adherents. One religion reporter spoke to a pastor who feared he was losing congregants to a cult—and like this pastor, evangelical leaders do not know what to do about the persistence of these bizarre, harmful ideas (see Kristian, 2020; Burke, 2020). There are many factors that come into the creation of the QAnon movement—from the mainstreaming of Russian disinformation tactics, to personal stories of abuse and mistrust of institutional authority—but the phenomenon can also be understood in light of the rapid media change of the moment. The gatekeepers that once successfully exercised control of information through broadcast media can be easily bypassed now. Charisma and mastery over digital tools have become ways for movements to gain steam, gain authority, and take over cultural spaces.

I have argued that digital habitus has changed evangelical culture, and QAnon may be another, darker example of this trend. As I have shown throughout the book, evangelicals have integrated new media technologies into nearly every aspect of evangelical culture. Just as evangelicals have historically used media technologies such as radio and television to spread their message, today's evangelicals take to digital media enthusiastically. From the digital church leaders who hope to reach a global population, to those entrepreneurs coding for a deeper purpose, to those missionaries who imagine the globe as a connected network of mobile phones, to those influencers and podcasters seeking new forms of community engagement and activism, evangelical culture has eagerly adopted the norms of the so-called digital age.

The case studies explored here also assert that evangelicals' shift to the digital is different in qualitative and fundamental ways from other shifts throughout their history. Although church leaders and traditional evangelical centers of authority have doubled down on using technology to appeal to the modern, American spiritual seeker, by catering to and promoting digital habitus they have also empowered the grassroots. And unlike with other media forms, the affordances of new media allow for ordinary users to become cultural authorities and to create publics, networks, and discourses that challenge the traditional evangelical power structure and the norms of this subculture. Evangelicals want to prove their relevance to the modern world by integrating technology into their culture, but in so doing, they have also let the modern world in. Thus evangelicalism has changed and is changing because of its embrace of digital habitus.

It may be that technology is a type of spiritual kryptonite that leads evangelicals into the delusional world of conspiracy, or it may be that it is the path that leads to a more robust Christian presence in the public sphere as most of the subjects in this book would hope. In either case, it seems impossible to imagine an American evangelicalism that completely divorces itself from the technological tools that have been thoroughly embedded into the ecosystem of this culture. Unlike other faith traditions, evangelicals have proven themselves ready to jump in and try out new media. And these experiments have changed what it means to worship, to proselytize, and to form communities in evangelical culture. It has never been my goal to say that evangelicals are right or wrong in their embrace of technology. My goal has been to show the complex effects consumer media technologies can have. And I have found that digital habitus has changed how evangelicals think and operate, and how they see themselves as inhabitants in the modern world and as passengers in Christian history.

If it is difficult to imagine an American evangelicalism that is not infused with techy-ness, it also may be because it is increasingly difficult to imagine being an American without consumer technologies. Americans are increasingly living lives in some digital realm or another, whether or not they are conscious of it. It is getting harder and harder to opt out of digital databases that store identity-defining metadata. We have been conditioned to understand that our photos and files are stored "in a cloud" somewhere. Like the evangelicals in this book, American culture and media have theologized the agentive or productively disruptive role that technology can play. Tech thinking has structured American thinking. And if there is a lesson that can be drawn from the case studies in these pages that might be applied to American culture as it struggles with the changes digital media has wrought, it is that media affects and changes us in complex ways that we cannot predict in advance and cannot reverse after the fact.

Many writers have used the metaphor of the mirror to describe our relationship to new media technologies. Like Narcissus, we are captivated by our own reflection and social media companies profit off of our vanity, allowing us more and more reflective surfaces on which we might play. And looked at on a large scale, this collective mirror held up to society is created and maintained by the digital habitus of its users and reveals things about the cultures we inhabit. Evangelicals looking into this mirror in the early days of the twenty-first century wanted to create a redeemed world. They thought that they could do this by planting churches, by influencing business culture, by spreading the gospel. But like a mirror, digital media also reveal blemishes, imperfections, things that need to be covered over, altered, changed. And the changes that have come have been swift, and to some, startling.

And so, it might be that if evangelical digital culture is like a mirror, it is a broken one that allows different people to see different things reflected. Digital habitus has changed the church in ways church leaders could not have predicted. Authoritative structures that have upheld the evangelical status quo are more easily weakened by unsanctioned voices like the popular evangelical influencers who have led the charge on changing the church's attitudes on sexism and abuse. There is more fracturing, more argumentation, more debate. And like Pandora's box, into the chaos comes misinformation and hate. Evangelicals have not redeemed the internet. QAnon is evidence of this fact. But it may be, many hope, that digital habitus redeems evangelicals by helping them face the historical inequities that have plagued their culture. If the digital world is a mirror—broken, funhouse, trick, dark, or black—evangelicals are in the process of recognizing themselves within it, and sometimes finding in their reflection hope for redemption.

Glossary

altar call   Typically occurring at the end of a sermon or other Christian event, this is the time when the pastor or orator urges anyone in the crowd to come to the front and "be saved."

The Body of Christ (sometimes abbreviated to "The Body")   The conceptualization of the global aggregate of all Christian believers in the world.

"capital C" Church   As in "The Body of Christ," this refers to all Christian believers in an ecumenical sense.

the Early Church   The Christian church described in the Book of Acts.

ex-vangelical (also exvangelical)   A former evangelical typically associated with online dissent against evangelical cultural norms.

fruit   This term comes from the Book of Matthew (7:15–20) that quotes Jesus as saying of false prophets "You will know them by their fruits." It refers to the idea that Christians should prove themselves based on their actions rather than their words. It is often used in reference to people's behavior.

fundamentalism   This term was inaugurated by a series of pamphlets called *The Fundamentals* in the early twentieth century. Fundamentalists advocated for a strict, literal understanding of the Bible and for Christians to retreat from worldly affairs. After the Scopes Trial of 1925, fundamentalism fell out of favor. Most American evangelicals do not consider themselves fundamentalists today.

gifts   This refers to an individual's spiritual strengths conceived of as springing from the Holy Spirit. Although it is meant to connote spiritual gifts, this term is also used more broadly as, for example, when a person is said to have "the gift of organization."

The Gospel   Specifically denotes the first four books of the New Testament but connotes what Christians see as the "good news" of the story of Jesus Christ and his significance as a savior of all people.

The Great Commission   This refers to the goal set by Jesus in the Book of Matthew (Matthew 28:16–20) that his followers go forth and make disciples of all people.

harvest   In the missionary context, this refers to the idea that there are souls in the world that need to be spiritually harvested, or saved, as a successful mission would reap a bountiful harvest.

Kingdom   Refers to the biblical idea of the Kingdom of God or the Kingdom of Heaven and is used to connote a spiritual world that is under the authority of God and Jesus.

Kingdom work   Work done in service to the Kingdom. Typically missions work and church work fall into this category.

Kingdom tools   Any technology that can be conceived of as helping with or doing "Kingdom work."

least of these   From the Book of Matthew (25:40), this refers to those who are poor, downtrodden or seen as less resourced.

missional   A way of being or doing things that is meant to attract people to Christianity.

parachurch   A Christian organization that is not a church, usually a nonprofit organization.

Pentecost   A biblical event that was said to have occurred after Jesus's ascension into Heaven. At this moment the Holy Spirit was said to have descended on Jesus's disciples. This is seen as the spiritual beginning of the Christian church. It is often referenced as a theological shorthand meant to signify that all Christian believers have the authority of the Holy Spirit.

revival   Usually meant to refer to a period in a church's history with a heightened spiritual significance. Revivals have been important events in American Christianity and their emotional, affective dimensions have been shaping forces in evangelicalism.

saved   The moment when a person commits to being a Christian by "accepting Jesus as a personal savior."

SBC   The Southern Baptist Convention, the largest evangelical organization in the United States.

spiritual formation   A person's spiritual growth path, or how a person grows in spirituality as a Christian, as in how a child in the church may grow up to become a faithful adult.

testimony   The way that a Christian may "witness" their faith, usually told as a personal conversion story.

the unchurched   Refers to all of those people who chose not to attend church. Usually does not refer to people in places where Christianity is not prevalent but rather references secular people in the United States.

unreached people groups   Coming from the missionary tradition, this refers to the theory that there are ethnolinguistic groups of people that cannot neatly be divided by nationality that have not been reached by Christianity.

walk, walks, walk with God   Often shortened to simply "walk," this refers to the proverbial "walk with God," that is every individual person's life path as a Christian. The idea of "different walks" signifies diverse life experiences.

witness, witnessing   "To bear witness as a Christian" is to share one's faith with others.

Word of God   Typically used to refer to the Bible.

Notes

1. Westboro has been denounced by many evangelical and mainline Christian denominations and institutions.

2. D. G. Hart (2004), for example, calls evangelicalism "the wax nose of twentieth-century Protestantism. Behind this proboscis, which has been nipped and tucked by savvy religious leaders, academics, and pollsters, is a face void of any discernible features" (p. 17).

3. This characterization of evangelicalism follows from David Bebbington's (1989) definition, which is also used by the National Association of Evangelicals and tends to be an accepted way to understand the foci around which evangelicals cohere.

4. See especially Balmer and Winner (2002), in which the authors offer a portrait of evangelicalism as a whole, and Hoover (2000), in which the author uses a visit to the famous Willow Creek megachurch in Illinois to explore how evangelical culture uses the semiotic register of popular culture to attract parishioners.

5. See especially Hangen (2002) on evangelicals' radio shows and their audiences, Hendershot (2004) on the material culture of evangelicalism, and Hoover (1988; 1990), Schultze (1991), Walton (2009), and Bowler (2013) on televangelism as a cultural form. As evangelicals have approached new media, Campbell's (2010) work has shown their continued interest in using media technologies and popular cultural forms to keep up with changes in American cultures.

6. See especially Kruse (2015), Dochuk (2011), Posner (2020), Butler (2021), and Fitzgerald (2017).

7. See Hendershot (2004).

8. As Hart (2004, p. 83) has noted: "After World War II until the 1970s, evangelicals constantly looked over their shoulders to see who was going to accuse them of fundamentalism."

9. Charles Fuller's *Old Fashioned Revival Hour* especially embodies the ethic of the neo-evangelical. Hangen (2002) has noted that although Fuller was a fundamentalist and a premillennialist—premillennialists believe that Christ will return to rapture his followers before the millennium prophesied in the Bible, the thousand-year reign of peace on Earth, and thus signs of the world falling into sinfulness and evil can be interpreted as signs of Jesus's imminent return—he toned down these themes in his broadcast and even claimed to be an apolitical figure. Hangen's (2002, p. 88) research shows that *The Old Fashioned Revival Hour* attracted a fervent national audience who communicated with Fuller and his wife through letters. The feeling of connection that these listeners had with the Fullers is evidence that the medium helped Fuller and other evangelicals inaugurate a new audience that could connect Christians around the country through the habitus of radio listening.

10. As evangelicals have moved online, they have also created new forms, such as the online church, which Tim Hutchings has ethnographically explored (2007; 2011; 2013). Relatedly, Robert Glenn Howard (2011) has pinpointed what he calls a new movement of evangelicals based on "vernacular Christian fundamentalism" that has come into being through digital sociality. Many evangelicals themselves have written books about the possibilities for social media (for some examples see Sweet [2012]; Stephenson [2011]; Murrell [2011]; Rice [2009]) and alternative versions of church like Second Life Churches (see Estes [2009]).

11. As Bourdieu has explained, "the habitus, the product of history, produces individual and collective practices, and hence history, in accordance with the schemes engendered by that history" (1977, p. 84)

12. Media Studies researchers, for example, have found evidence that greater internet skills positively affect the subjective well-being of older adults. See Hofer et al. (2019).

13. For example, Alice Marwick's (2013) ethnography of Silicon Valley argued that Silicon Valley culture has internalized the myth that it is a meritocratic industry. This cultural understanding has excluded women and people of color because it is assumed that if they do not already have a seat at the table, they must not have earned one—and may not deserve one. And Sarah Wachter-Boettcher (2017) has made a similar point, more bluntly writing: "Scratch the surface at all kinds of companies—from Silicon Valley's 'unicorns' (startups with valuations of more than a billion dollars) to tech firms in cities around the world—and you'll find a culture that routinely excludes anyone who's not young, white, and male" (p. 16).

14. As Zeynep Tufekci (2017) has described it, the digital media technologies are powerful tools for social movements but reliance on them can also attenuate social movements in the long run because they do not allow for the on-the-ground mobilizing and organizing that social movements need to survive and thrive.

CHAPTER 1. THE CHURCH

1. See Pew Research Center (2015) on this trend.

2. Life.Church was formerly called LifeChurch.tv. They changed their name in 2015.

3. This chapter is based on a variety of qualitative sources. I conducted semi-formal interviews with church leaders: communications directors, pastors, and others; attended two church conferences, one in Dallas in the summer of 2013 and one virtual conference in the summer of 2017; I analyzed twenty-six books written by evangelicals on how churches should approach the digital era; I attended evangelical church services in Los Angeles, New York City, San Francisco, New Jersey, Dallas, and Nashville, and spent three days at the church that provides the central data for this chapter, Life.Church in Edmond, Oklahoma. At two Life.Church locations in Edmond and at their central offices I conducted roughly thirty informal interviews with parishioners, volunteers, and staff members (pseudonyms are used for the names of all informal interviews from Life.Church). I argue that the hybridized discourse growing from the megachurch movement that saw the twinning of business speak with Christian strategies and now takes cues from the world of high technology has fundamentally changed the central institutions that define evangelicalism today.

4. The Echo Conference ran from 2007 through 2013.

5. In her ethnography of hackers, Gabriella Coleman (2013, p. 47) has theorized the conference as a social ritual that reinforces group solidarity. Although church conferences are different than hacker conferences in that they do not include the element of making as a central activity, they have a ritualistic aspect that is heightened, I found, by the ritualistic activity common to evangelical gatherings, for example, prayer—which at Echo often occurred before or after a conference presentation.

6. There is an extensive body of scholarly literature on the American spiritual marketplace. See especially Ellingson (2007) on the megachurch and the spiritual marketplace and Banet-Weiser (2012, chapter 5), whose work looks at how Christian preachers have mobilized an understanding of branding and marketing to attract spiritual seekers.

7. See also Wade Clark Roof's (1993) book, *A Generation of Seekers.*

8. Kevin Kruse (2005) has theorized that the "white flight" that defined Atlanta, Georgia, was the "politics of suburban secession" (p. 234) and was always

already a racialized project. Darren Dochuk (2011) has connected white flight to the birth of contemporary evangelical politics. The racial integration of schools was the first one of the first issues that galvanized and connected the network of people who would go on to become the Christian Right. As mass migration filled the suburbs of the Golden State with transplants from the South and these "Christian citizens living west of the Mississippi believed their true calling was to advance the Christian heritage passed down to them from their Anglo-Saxon or Scotch-Irish ancestors" (Dochuk, 2011, p. 13), California became the epicenter of white evangelical conservatism.

9. Emerson and Smith (2001) have explained how evangelicals understand this evident fact, but often brush it off. They write, characterizing the white evangelical response: "Although it is perhaps not the ideal case, there is certainly nothing wrong with attending racially distinct congregations, as long as the motivation is not prejudice. People are comfortable with different worship styles, want to be with familiar people, and have different expectations about congregations. For these reasons, if congregations end up being racially homogenous, it is acceptable, if not preferable" (p. 186). Although of course there are exceptions. There are many Black, Asian, and Latino church planters who have successfully grown large, thriving church communities. And there are those pastors who have self-consciously and carefully tried to cultivate an ethically and racially diverse congregation—for example, Erwin McManus of Mosaic Church in Los Angeles, which has been written about by Gerardo Marti (2005).

10. Research from the Hartford Institute (Bird & Thumma, 2020) found that 70 percent of megachurches in 2020 were multisite churches. In contrast, in 2010 only 46 percent of megachurches were multisite. As this data indicates, multisite churches are a growing trend.

11. For a succinct history of the many forms that online churches have taken, see Hutchings (2017), chapter 1, "A brief history of cyberchurch," pp. 10–23. And for a history of Life.Church's online church and the various changes it has undergone it its history, see Hutchings (2017), chapter 8, "Church online at LifeChurch.tv."

12. See YouVersion (n.d.), which tracks downloads of the Bible App.

13. In other words, online church is not a "virtual world," as Tom Boellstorff's (2008) definition would have it. Boellstorff explains that to be a "virtual world," an online platform must meet three requirements. First, it must be a place with a sense of placeness. In the game *Second Life*, for example, residents buy "land" and understand themselves in relation to a spatial reality created by the world. Second, for a site to be a virtual world, it has to have people in it. Thus it has to be a site of sociality. Third, and perhaps most obviously, it has to be online. These three requirements paint a specific picture of virtual worlds as online "places" in which a social world is performed. There is some evidence that proponents of online church hope that attending a church online feels like entering a place, and indeed a social world.

CHAPTER 2. THE START-UP

1. Interviews were conducted in person in New York City, Silicon Valley, Los Angeles, and Nashville and via Skype, Zoom, or phone with those start-ups located in other places in the country and in one case in Australia. I interviewed two of my informants more than once and kept in touch with many of them via Twitter and via the apps that they had created. In this chapter, and in the chapters that follow, I cite interviews when those I interviewed agreed to be named in this book. Those who chose to remain anonymous have been given pseudonyms. When I use pseudonyms in the text I indicate that these are not real names.

2. Thomas Streeter (2011) has noted that Steve Jobs especially fit the bill of the 1980s entrepreneur and his iconoclasm only served to solidify his place in "the mythic American narrative of the entrepreneur, who in popular fantasy came from nowhere and needed no outside support" (p. 69).

3. The dazzling financial success of these self-made men proved to many that the right-wing economic policies promoted by Reagan and Thatcher in the 1980s were justified. As Streeter has explained, "The microcomputer thus provided a sophisticated, high-tech glitter to the Reagan era enthusiasm for markets, deregulation, and free enterprise; it became an icon that stood for what's good about the market, giving leaders the world over an extra incentive to pursue neoliberal policies" (2011, p. 87).

4. See Jenkins (2006) and Benkler (2006) for examples of academic texts that had optimistic views of the internet's democratic potential.

5. See Schulte (2013, pp. 139–163).

6. See Merchant (2017, chapter 13 "Sellphone").

7. See Stewart (2013).

8. I caught up with Dean Sweetman in 2021 and he told me that this figure—85 percent of donations to churches are in cash—had likely changed to around 65 percent four years later due to the development of apps like his and other technologies.

CHAPTER 3. MEDIA MISSIONS

1. See McAlister (2019, chapter 11) for an analysis of the role of short-term missions work and organizations.

2. See, for example, McAlister (2019, chapter 1).

3. See Pratt (2016) and McAlister (2019).

4. See Fitzgerald (2017, pp. 479–483).

5. David Nord (2004) has argued that the missionary impulse among early American Christians led to the founding of tract societies across the United States, and this was the nation's first mass media.

6. See Kenny & Sandefur (2013, p. 75).

7. See Berry (2008, p. 101).

8. Although two of my informants suggested that Open may have put for-profit faith-tech platforms out of business, their stories are anecdotal. I have not found evidence proving that this is the case.

9. See Boyle (2008, p. 182).

10. See especially Balmer (1989, pp. 193–208).

11. In other words we have to get away from the technology industry's understanding of "edge cases." Sarah Wachter-Boettcher (2017) has explained that the technology industry, especially given its demographic makeup as a male-controlled space, has blind spots that influence how they imagine technology use. Everything outside of the frame that they set up is considered an "edge case." This is problematic given that the technology industry is concentrated in a relatively few areas, and there is very little input taken into consideration from people who inhabit "alternate media worlds."

12. Daniel Miller and Don Slater (2001) wrote in their early ethnography of internet use in Trinidad that Trinidads saw the internet as "naturally Trini." Miller and Slater's intervention, however, was to explain that internet use cannot be understood as distinct from the cultural practices of places—that the internet has to be studied in the real places and cultures in which it is used.

CHAPTER 4. THE INFLUENCERS

1. In this chapter I have decided not to cite Twitter or social media users unless they are recognized public figures. Although I have logged those tweets that I do not cite that I use in my analysis in my own records, I felt that citing them publicly was not ethical. I made the demarcation between recognized public figures and ordinary people by paying attention to the markers used by social media companies—for example, the blue check mark systems on on Twitter and Instagram. The ethics of social media research are complicated. Although tweets are acts of public speech, and Twitter users who tweet in response to hashtags display a willingness to involve themselves in public conversation, this act in itself does not equal their consent as research participants. Furthermore, women who come forward on a hashtag to share their stories of abuse may not expect or anticipate that those stories might become fodder for a research project. Feminist social media researchers have developed research practices that take these concerns into account. Moya Bailey (2015), for example, has mapped out a framework for "collaborative consent" in network research, and while my research does not model this ideal, my decisions have been inspired by the feminist ethos that requires that researchers go beyond the typical standards of an Institutional Review Board (IRB) in favor of a more considered, more robust ethical practice.

2. Brooke Duffy and Jefferson Pooley (2019) have called the celebrities of the social media age "idols of promotion," updating Leo Lowenthal's understanding of

celebrity from the 1940s, because of the way they are characterized and characterize themselves through their success at self-branding.

3. See, for example, Rachel Monroe's (2017) exploration of the hashtag #Van-
Life and the microcelebrities that populate it.

4. As Alice Marwick (2013) has written, "authenticity"—though a slippery, often
ill-defined term—is one of the guiding principles of social media presentation. In
her study of the culture of Silicon Valley entrepreneurship, Marwick has discussed
how the cultivation of authenticity is central for the maintenance of a self-brand,
and in turn a successful brand is the most powerful commodity in the social media
world—a commodity that she notes also has real economic value (2013, p. 167). An
understanding of "the authentic" also has religious value in evangelical communities. Deborah Whitehead's (2015) study of evangelical "mommy bloggers" argues
for the importance of understanding how the rhetoric of "authenticity" shapes
online publics in the Christian media sphere, and she writes that "the trust that
committed readers have in a blogger and her story seems in many ways to resemble an evangelical model of religious belief, relying on faith as 'the evidence of
things unseen'" (p. 141). Thus evangelicals are particularly primed to understand
"authenticity" as a central value in the online sphere.

5. See Hubner (2015), Perry (2013), Colaner & Giles (2008).

6. Although complementarians tend to reject feminism qua feminism, it would
be a mistake to assume that women in complementarian denominations are cut off
from the vagaries of popular and academic feminist discourse. C. Manning (1999)
found in her qualitative research with conservative women that most agree with
many of the values of feminism; they tend to enjoy working outside of the home,
expect fairness in their workplace, and hold positions of authority in their families. And though complementarianism in some ways rigidly fixes gender roles, in
her ethnographic account of two fundamentalist Christian congregations, Brenda
Brasher (1998) has theorized that the complementarian church is a "sacred canopy
with a sacred partition," meaning that women occupy enclaved spaces and exercise
their authority from within these strictures. Ethnographic and qualitative studies
focusing on evangelical women paint a more complex picture than the simple split
between complementarian and egalitarian theology allows; these studies find that
though men are in positions of nominal authority, women find ways to manipulate
the structure in which they find themselves to create space for their own voices.

7. See Katie Bowler (2019) on female celebrities and in particular the figure of
the preacher's wife. Anthea Butler (2012) has written about the role of the "church
mother" in the Church of God in Christ, a primarily Black Pentecostal-Holiness
denomination. Butler explains that church mothers have been the guardians of
tradition and collective memory, noting that their role carries as much importance
and authority as that of the male preacher in the church even when it seems nomi-
nally or institutionally less valid.

8. For example, another evangelical celebrity, Christine Caine, from the über-

popular Australian church and Christian media center Hillsong, established Propel, an organization whose mission statement reads: "Propel exists to see every woman activated in order to fulfill her God-given destiny" (Propel Women, 2021). See also Kate Shellnutt's (2014) article for more information on Propel.

9. See Kristin Kobes Du Mez (2020, p. 199) in which she discusses the "smokin' hot wife" phenomenon in evangelical culture and connects it to discourses of masculinity.

10. The surprising loyalty that evangelicals have for Donald Trump is detailed in Posner (2020), which outlines the NeverTrump movement in evangelical culture and its subsequent fallout after white evangelicals voted overwhelmingly for Trump.

11. For example, the 2012 Resolution on Same Sex Marriage from the Southern Baptist Convention states: "We express our love to those who struggle with same-sex attraction and who are engaged in the homosexual lifestyle; and ... we encourage our fellow Southern Baptists to consider how they and their churches might engage in compassionate, redemptive ministry to those who struggle with homosexuality" (Southern Baptist Convention, 2012).

12. See, for example, Dreher (2020).

13. Much of this discussion was previously published as Laughlin (2020).

14. As Pamela Klassen and Kathryn Lofton (2013) have noted: "Christian women have been eager to find media of witness to extend their specific embodiment beyond themselves and thus to make a mission of their particular experience" (p. 54).

15. It is important to note that the feminist movement I am describing here shares little in common with the "pit-bull feminism" embodied most famously by Sarah Palin and that saw evangelical antifeminist women hope to attain positions of power in order to dismantle any feminist gains (see Butler, 2012; Douglas, 2010, pp. 267–297; McCarver, 2012; Rodino-Colocino, 2012). It is rather inspired and fueled by female Christian writers and influencers, to promote a diverse array women's voices in a patriarchal subculture.

16. "Postfeminism" is a contested term with sometimes conflicting referents (see Gill 2007, 249–272). Here, I am using it as Rosalind Gill has theorized—as a sensibility marked by certain predilections present in media and popular culture that is "organized around notions of choice, empowerment, self-surveillance, and sexual difference and articulated in an ironic and knowing register in which feminism is simultaneously taken for granted and repudiated" (2007, p. 271). This sensibility relies on an understanding that many of the struggles of second-wave feminism have been overcome and thus feminism is no longer needed and takes as proof-positive the many individual women who have succeeded in their careers and in governments. As Sarah Banet-Weiser (2012) has put it: "The individual entrepreneur becomes the signature of a postfeminist women" (p. 61). Thus it is a sensibility that relies on a neoliberal understanding of the economic and politi-

cal world and regards with irony any attempt to critique gender inequities. For another theorist of postfeminism, Angela McRobbie (2009), this is the result of an insidious process that has in effect neutered potentially radical female critique. For McRobbie, postfeminism is a prescribed form of feminism that only allows women to occupy particular cultural spaces and does not allow for bonds of female solidarity to form.

17. See Klassen & Lofton (2013, pp. 60–62) on the career of Ann Voskamp.

18. Of course social media is not a purely democratic space as many scholars have pointed out. See, for example, Sunstein (2017); Poell & van Dijck (2018).

19. This is particularly significant given that many theorists have argued that postfeminism is an ethos that prizes individualism and individual achievement above all. It is a neoliberal understanding of the world that does not allow for collective engagement to take place. Angela McRobbie has called theorized this with regard to the cultural politics of disarticulation (2009, pp. 24–53). She draws on Stuart Hall's understanding of "articulation" in politics—the idea that in a deliberative democracy political power comes from the ability to draw connections among multiple subjectivities—to make the case that feminism has been disarticulated. She writes: "In social and cultural life there is instead a process of unpicking the seams of connection, forcing apart and dispersing subordinate social groups who might have possibly found some common cause" (McRobbie, 2009, p. 26). For McRobbie, feminism has been disarticulated from antiracism, and thus it is significant to see this boundary maintained.

20. See Smietana (2018) for an in-depth description of the Hybels case.

21. See Smietana (2021).

CHAPTER 5. RACIAL RECKONING AND REPAIR

1. See especially Marti (2005).

2. I focused on podcasts that had a history of more than a year and that had become somewhat institutionalized in that they were connected to Christian non-profits, parachurch organizations, and or conferences. Because anyone can have a podcast, finding the right sample to analyze is difficult: Which podcasts have listeners? Which are truly representative of the moment? How do we judge that? I chose the metrics of audience engagement and longevity. I steered clear of radio shows that repurpose their content as podcasts. My reasoning for this is that I wanted to understand what was going on off of the networks rather than what is sanctioned by network executives. I chose three podcasts that were all slightly different in format. *Pass the Mic* could be considered a "preacher podcast" in which the hosts presented ideas and themes in the manner and using the rhetorical styles of preaching. *The Jude 3 Project* is a podcast focused on Christian apologetics. *The Church Politics Podcast* is a policy focused podcast that analyzes politics from a Christian perspective. I performed a discourse analysis of three podcasts (*Pass the*

Mic, The Jude 3 Project, and *The Church Politics Podcast)* that were posted from March 15 through September 15, 2020—a six-month period. I listened to about forty-five hours of tape and coded the podcasts for themes using a Grounded Theory (see Charmaz, 2014) approach.

3. See Park & Davidson (2020) for a discussion of the political polling as it relates to the reification of white Christianity.

4. This quote comes from James Baldwin's unpublished writings that were used in the 2016 documentary *I Am Not Your Negro* directed by Raoul Peck.

5. See Emerson & Smith (2001, pp. 38–42); Tisby (2019, chapters 4 and 5); and Marti (2020, chapters 2 and 3).

6. For example, the famous revivalist preacher Charles Finney, a leader in the abolition movement desecrated Oberlin College in the 1840s when he was the president of that institution (see Hambrick-Stowe, 1996).

7. I use the umbrella term "the Black church" in this chapter to refer to the many varied denominations and churches that boast primarily African American congregants and tend to share liturgical and theological norms. This was the term that my informants gave to the movement, and it has emic value for that reason. The idea of "the Black church" as the religious and cultural phenomenon we understand it to be today was first laid out by the Black sociologist E. Franklin Frazier (1964). See also Lincoln & Mamiya (1990) for a more complete understanding of the theological and denominational intricacies of this movement. And see Nelsen & Nelsen (1975) for a history of the Black church in the twentieth century and Billingsley (1999), whose work takes an ethnographic perspective of the Black church in urban and rural contexts.

8. See Collins (2012, pp. 82–83) and Butler (2021, chapter 2).

9. See Pimblott (2016) on how the discourses that merged Black power and a Christian worldview provided the discursive background of the Civil Rights movement. See also Anthony Pinn (2002), who has argued that the Black church was central in activist organizing during this time, because "in addition to providing bodies willing to participate in direct action, disseminate information, and finance protest activities, The Black Church also provided the ideology and theological underpinning for the movement" (p. 13).

10. See Bartowski (2004), Abraham (1997), and Du Mez (2020, chapter 9, "Tender Warriors") for more on the Promise Keepers movement.

11. See Abraham (1997, chapter 8, "Are Promise Keepers Racist?").

12. See Fitzgerald (2017, p. 617) and Posner (2020, chapter 3).

13. See Fea (2018) on the rise of Trump's "court evangelicals."

14. See Robertson (2018).

15. This figure has been reported by Onishi (2019).

16. See Vincent Meserko (2015) in which he analyzes Marc Maron's popular podcast and explores how the podcast format lends itself to authentic expressions of selfhood.

17. See Phillips (2015, chapter 6).

18. See Marwick & Boyd (2012), especially page 120 in which the authors explain that "users write different tweets to target different people (e.g., audiences). This approach acknowledges multiplicity, but rather than creating entirely separate, discrete audiences through the use of multiple identities or accounts, users address multiple audiences through a single account, conscious of potential overlap among their audiences."

19. For example, Lifeway Christian Resources publishing made the decision to pull Jen Hatmaker's books from their shelves when she came out in support of gay Christian identity, as I discussed in chapter 4.

20. See https://www.thegospelcoalition.org/ for more on the Gospel Coalition and their annual conference.

21. Michael Wear left his position as podcast cohost in the fall of 2020 and was replaced by Chris Butler.

22. See Kimberlé Crenshaw (1989), who is widely understood to have coined the term "intersectionality," and Patricia Hill Collins (2019), who has written about the legacy of intersectionality and the potential for it to be a critical social theory.

23. *Truth's Table* tweet (2017, April 15), https://twitter.com/TruthsTable/status/853217014764240896.

24. Todd Pruitt's response, "Old Error for a New Generation," is at his blog, *Alliance of Confessing Evangelicals*, https://www.alliancenet.org/seen-and-heard/old-error-for-a-new-generation.

References

Abraham, K. (1997). *Who are the Promise Keepers? Understanding the Christian men's movement.* Doubleday.

Arulkumarasan, J. (2017). Personal interview. Skype.

Ames, M. (2019). *The charisma machine: The life, death, and legacy of One Laptop Per Child.* MIT Press.

Bailey, M. (2015). "#transform(ing)DH writing and research: An autoethnography of digital humanities and feminist ethics." *Digital Humanities Quarterly, 9*(2). http://www.digitalhumanities.org/dhq/vol/9/2/000209/000209.html

Balmer, R. (1989). *Mine eyes have seen the glory: A journey into the evangelical subculture in America.* Oxford University Press.

Balmer, R., & Winner, L. F. (2002). *Protestantism in America.* Columbia University Press.

Banet-Weiser, S. (2012). *Authentic™: The politics of ambivalence in a brand culture.* New York University Press.

Banet-Weiser, S. (2018). *Empowered: Popular feminism and popular misogyny.* Duke University Press.

Barbrook, R., & Cameron, A. (1996). The Californian Ideology. *Science as Culture, 6*(1), 44–72.

Bartkowski, J. P. (2004). *The Promise Keepers: Servants, soldiers, and godly men.* Rutgers University Press.

Bebbington, D. W. (1989). *Evangelicalism in modern Britain: A history from the 1730s to the 1980s.* Unwin Hyman.

Benkler, Y. (2006). *The wealth of networks: How social production transforms markets and freedom.* Yale University Press.

Berg, T. C. (2003). Copying for religious reasons: A comment on principles of copyright and religious freedom. *Cardozo Arts & Entertainment Law Journal, 21*(2/3), 287–317.

Berry, D. (2008). *Copy, rip, burn: The politics of copyleft and open source.* Pluto Press.

Bessey, S. (2013). *Jesus feminist: An invitation to revisit the Bible's view of women.* Howard Books.

Billingsley, A. (1999). *Mighty like a river: The Black church and social reform.* Oxford University Press.

Bird, W., & Thumma, S. (2020). Megachurch 2020: The changing reality in America's largest churches. [Research report]. The Hartford Institute.

Blanchard, D. (2014a, August 15). An alternative imagination. https://medium .com/@dave_blanchard/an-alternative-imagination-6a59e1071fc5

Blanchard, D. (2014b, August 20). Frontiers of faith and work. https://medium .com/@dave_blanchard/frontiers-of-faith-work-a01113b7086c

Boellstorff, Tom. (2008). *Coming of age in second life: An anthropologist explores the virtually human.* Princeton University Press.

Bourdieu, P. (1965/1990). *Photography: A middle-brow art.* Stanford University Press.

Bourdieu, P. (1977). *Outline of a theory of practice.* Cambridge University Press.

Boyle, J. (2008). *The public domain: Enclosing the commons of the mind.* Yale University Press.

Bowler, K. (2013). *Blessed: A history of the American prosperity gospel.* Oxford University Press.

Bowler, K. (2019). *The preacher's wife: The precarious power of evangelical women celebrities.* Princeton University Press. doi:10.2307/j.ctvfjd055

Bracey, G. E., & Moore, W. L. (2017). "Race Tests": Racial boundary maintenance in white evangelical churches. *Sociological Inquiry, 87*(2), 282–302.

Brasher, B. (1998). *Godly women: Fundamentalism and female power.* Rutgers University Press.

Brock, A. (2020). *Distributed Blackness: African American cybercultures.* New York University Press.

Brown, A. C. [@austinchanning]. (2017, April 27). We survived the printing press, radio, televangelists ... I think we will survive the blogosphere, and whatever is next. I'm not worried [Tweet]. Twitter.com. https://twitter.com /austinchanning/status/857779935221559297

Brown, A. C. (2020). *I'm still here: Black dignity in a world made for whiteness.* Penguin Random House.

Bryant-Davis, T. (2020). Personal interview. Phone.

Buice, J. (2019, May 28). Why the SBC should say "no more" to Beth Moore.

Delivered by Grace [Blog]. https://www.deliveredbygrace.com/why-the-sbc
-should-say-no-more-to-beth-moore/

Burke, D. (2020, October 15). How QAnon uses religion to lure unsuspecting
Christians. CNN. https://www.cnn.com/2020/10/15/us/qanon-religion-churc
hes/index.html

Burns, T. (2020). Personal interview. Zoom.

Butler, A. D. (2012). *Women in the Church of God in Christ: Making a sanctified
world*. University of North Carolina Press.

Butler, A. D. (2021). *White evangelical racism*. University of North Carolina
Press.

Campbell, H. (2010). *Where religion meets new media*. Routledge.

Carlson, D. (2020, October 6). The white savior complex in missions?: And what
it reveals about us. *The Gospel Coalition*. https://www.thegospelcoalition.org
/article/white-savior-complex-missions/

Carrette, J., & King, R. (2005). *Selling spirituality: The silent takeover of
religion*. Routledge.

Case, J. R. (2012). *An unpredictable gospel: American evangelicals and world
Christianity 1812–1920*. Oxford University Press.

Charmaz, K. (2014). *Constructing grounded theory*. Sage.

Chastain, B. (Host). (n.d.). *Exvangelical* [Audio podcast]. https://www.exvange
licalpodcast.com/

Clark, H. (2020, June 29). Apostate author Jen hatmaker reveals her daughter is
lesbian in honor of pride month. *Christian News*. https://christiannews.net
/2020/06/29/apostate-author-jen-hatmaker-reveals-her-daughter-is-lesbian
-in-honor-of-pride-month-im-so-glad-youre-gay/

Code for the Kingdom. (2017). Code for the kingdom. [webpage]. https://codefo
rthekingdom.org/

Colaner, C. W., & Giles, S. M. (2008). The baby blanket or the briefcase: The
impact of evangelical gender role ideologies on career and mothering aspira-
tions of female evangelical college students. *Sex Roles 58*(7/8), 526–534.

Coleman, G. (2013).*Coding freedom: The ethics and aesthetics of hacking*.
Princeton University Press.

Collins, K. (2012). *Power, politics and the fragmentation of evangelicalism: From
the Scopes Trial to the Obama administration*. InterVarsity Press.

Comcast/ Xfinity. (2014). The future of awesome [Television advertisement].
https://www.ispot.tv/ad/7RRN/xfinity-internet-tech-startup

Cone, J. (1969/2018). *Black theology and Black power*. Orbis Books.

Council on Biblical Manhood and Womanhood. (2017). The Nashville Statement.
https://cbmw.org/nashville-statement/

Covenant Eyes (2020). Home page. [Website]. https://www.covenanteyes.com/

Crenshaw, K. (1989). Demarginalizing the intersection of race and sex: A Black feminist critique of antidiscrimination doctrine, feminist theory, and antiracist politics. *University of Chicago Legal Forum 1*(8), 140–167.

Cybermissions.org. (n.d.). https://www.cybermissions.org/.

Dochuk, D. (2011). *From Bible Belt to Sunbelt: Plain-folk religion, grassroots politics, and the rise of evangelical conservatism* . Norton.

Douglas, S. (2010). *The rise of enlightened sexism: How pop culture took us from girl power to girls gone wild*. St. Martin's.

Dreher, R. (2017, December 18). Of Hatmaker and heresy. *The American Conservative*. https://www.theamericanconservative.com/dreher/jen-hatmaker-he resy-evangelical/

Dreher, R. (2020 January 10). The insanity of transgenderism. *The American Conservative*. https://www.theamericanconservative.com/dreher/the-insanity -of-transgenderism/

DuBois, W.E.B. (1897). Strivings of the Negro people. *The Atlantic Magazine*.

Du Mez K. K. (2020). *Jesus and John Wayne: How white evangelicals corrupted a faith and fractured a nation*. Liveright.

Duffy, B. E., & Pooley, J. (2019). Idols of promotion: The triumph of self-branding in an age of precarity. *Journal of Communication, 69*(1), 26–48.

EchoHub (2018). Thank you for 6 amazing years. [Website]. http://echohub.com

Edmiston, J. (2017). Personal interview. Torrance, CA.

Ellingson, S. (2007). *The megachurch and the mainline: Remaking religious tradition in the twenty-first century*. University of Chicago Press.

Emerson, M. O., & Smith, C. (2001). *Divided by faith: Evangelical religion and the problem of race in the church*. Oxford University Press.

Estes, D. (2009). *SimChurch: Being the church in the virtual world*. Zondervan.

Evans, C. (2018). *Broadband: The untold story of the women who made the internet*. Portfolio/Penguin.

Evans, R. H. (2015). *Searching for Sunday: Loving, leaving, and finding the church*. Nelson Books.

Evans, R. H. [@rachelheldevans]. (2017, April 27). LITERALLY THREW MY PHONE ACROSS THE BEDROOM OVER THIS PIECE [Tweet]. Twitter. com. https://twitter.com/rachelheldevans/status/857638725173600261

Fanon, F. (1963/2004). *The wretched of the earth*. Grove Press.

Fea, J. (2018). *Believe me : The evangelical road to Donald Trump*. Eerdmans.

Feddes, M. (2012, April 20). Richard Land's comments on Trayvon Martin investigated by SBC. *Christianity Today*. https://www.christianitytoday.com/news /2012/april/richard-lands-comments-on-trayvon-martin-investigated-by.html

Fields, L. (Host). (2015–present). *The Jude 3 Project* [Podcast].

Fields, L. (Host). (2020a, April 25). African theology–special guest Vince Bantu [Podcast episode]. *The Jude 3 Project* [Podcast].

Fields, L. (Host). (2020b, July 14). Listen, lament, legislate [Podcast episode]. *The Jude 3 Project* [Podcast].

Fitzgerald, F. (2017). *The evangelicals: The struggle to shape America*. Simon and Schuster.

Florini, S. (2015). The podcast "Chitlin Circuit": Black podcasters, alternative media, and audio enclaves. *Journal of Radio & Audio Media*, 22(2), 209–219.

Frazier, E. F. (1964). *The Negro church in America*. Schocken Books.

Frederick, M. (2003). *Between Sundays: Black women and everyday struggles of faith*. University of California Press.

Giboney, J., & Wear, M. (Hosts). (2020, June 10). Civic update—action steps to equip the church to fight racial injustice [Podcast episode]. *The Church Politics Podcast* [Podcast].

Gill, R. (2007). *Gender and the media*. Polity Press.

Gilson, R. (2020). Coming out and coming to faith. [Blog post]. https://www.rac helgilson.com/blog-index/coming-out-and-coming-to-faith

Ginsburg, F. (2008). Rethinking the digital age. In D. Hesmondhalgh & J. Toynbee (Eds.), *The Media and Social Theory* (pp. 127–144). Routledge.

Glick, P., & Fiske, S. (1996). The ambivalent sexism inventory: Differentiating hostile and benevolent sexism. *Journal of Personality & Social Psychology*, 70(3), 491–512.

Gormly, E. (2003). Evangelizing through appropriation: Toward a cultural theory on the growth of Contemporary Christian Music. *Journal of Media and Religion*, 2(4), 251–265.

Green, E. (2017). A conservative Christian battle over gender. *The Atlantic Magazine*. https://www.theatlantic.com/politics/archive/2017/07/truths-table -gender-race/532407/

Green, E. (2018). The tiny blond Bible teacher taking on the evangelical political machine. *The Atlantic Magazine*. https://www.theatlantic.com/magazine/arc hive/2018/10/beth-moore-bible-study/568288/

Groeschel, C. (2007). Open source ministry. *Ministry Today*. http://ministryto daymag.com/index.php/ministry-today-archives/66-unorganized/14356-open -source-ministry

Habermas, J. (1962/1989). *The structural transformation of the public sphere: An inquiry into a category of bourgeois society*. MIT Press.

Hambrick-Stowe, C. (1996). *Charles G. Finney and the spirit of American evangelicalism*. Eerdmans Publishing.

Hangen, T. J. (2002). *Redeeming the dial: Radio, religion, and popular culture in America*. University of North Carolina Press.

Hart, D. G. (2004). *Deconstructing evangelicalism: Conservative Protestantism in the age of Billy Graham*. Baker Academic.

Hatmaker, J. (2016, April 23). *After our beautiful, beautiful event today* [Status update]. Facebook. https://www.facebook.com/permalink.php?story_fbid=94 6752262090436&id=203920953040241

Hatmaker, J. (2017a). *Of Mess and Moxie: Wrangling delight out of this wild and glorious life*. Thomas Nelson.

Hatmaker, J. (2017b). About me. [Website]. https://jenhatmaker.com/about/

Hatmaker, J. (2017c, August 29). The fruit of the "Nashville Statement" is suffering [Tweet]. https://twitter.com/jenhatmaker/status/902590545578643456?lang=en

Hatmaker, J. (Host). (2020, June 26). A moment of pride: Jen and Sydney Hatmaker on being gay and loved (No.6) [Audio podcast episode]. *For the love* [Podcast]. https://jenhatmaker.com/podcast/special-edition-series/a-moment-of-pride-jen-and-sydney-hatmaker-on-being-gay-christian-and-loved/

Hendershot, H. (2004). *Shaking the world for Jesus: Media and conservative evangelical culture*. University of Chicago Press.

Henny, A. (2020). Personal Interview. Zoom.

Higgins, M., Edmondson, C., & Uwan, E. (2017, March 25). Gender apartheid [Podcast episode]. *Truth's Table* [Podcast].

Hill Collins, P. (2000). *Black feminist thought*. Routledge.

Hill Collins, P. (2019). *Intersectionality as critical social theory*. Duke University Press.

Hofer M., Hargittai, E., Buchi, M., & Seifert, A. (2019). Older adults' online information seeking and subjective well-being: The moderating role of internet skills. *International Journal of Communication (19328036)*, *13*, 4426–4443.

Hollandsworth, S. (2018). The power issue: Beth Moore is forcing evangelical Christianity to get woke. *Texas Monthly*. https://www.texasmonthly.com/the-culture/power-issue-beth-moore-forcing-evangelical-christianity-get-woke/

Hollis, R. [@msrachelhollis]. (2015 March 21). I have stretch marks and I wear a bikini. I have a belly that's permanently flabby from carrying three giant babies and I [Instagram post]. Instagram. https://www.instagram.com/p/ofwr9Dl-VK/?hl=en

Hollis, R. [@msrachelhollis]. (2020 November 21). Dropped into LA for just long enough to get these roots in line. Now it's back home to my babes [Instagram post]. Instagram. https://www.instagram.com/p/CH33BXAJpVk/c/17849932109567621/

Hollis, R. (2018). *Girl, Wash Your Face*. Thomas Nelson Books.

Hollis, R. (2019). *Girl, stop apologizing: A sham-free plan for embracing and achieving your goals*. HarperCollins.

Hollis, R. (2020). *Didn't see that coming: Putting life back together when your world falls apart*. Dey Street Books.

Holmes Jr., C. (2016, June 2). Urban church planting in a ministry world. *The Witness*. https://thewitnessbcc.com/urban-church-planting/

Hoover, S. M. (1988). *Mass media religion: The social sources of the electronic church*. Sage Publications.

Hoover, S. M. (1990). The meaning of religious television. In Q. Schultze (Ed.), *American evangelicals and the mass media* (pp. 231–251). Zondervan.

Hoover, S. M. (2000). The cross at Willow Creek: Seeker religion and the contemporary marketplace. In B. D. Forbes and J. H. Mahan (Eds.), *Religion and popular culture in America* (pp. 145–160). University of California Press.

Horton, D., & Wohl, R. R. (1956). Mass communication and para-social interaction; observations on intimacy at a distance. *Psychiatry, 19*(3), 215–229.

Howard, R. G. (2011). *Digital Jesus: The making of a new Christian fundamentalist community online*. New York University Press.

Hübner, J. (2015). The evolution of complementarian exegesis. *Priscilla Papers 29*(1), 11–13.

Hunter, J. C. (2007). *Church distributed: How the church can thrive in the coming era of connection*. Distributed Church Press.

Hutchings, T. (2017). *Creating church online: Ritual, community, and new media*. Routledge.

Indigitous. (n.d.). About us. https://indigitous.org/about-us/

Jackson, J. L. (2010). On ethnographic sincerity. *Current Anthropology*, 51S279–S287.

Jackson, S. J., & Foucault Welles, B. (2016). #Ferguson is everywhere: Initiators in emerging counterpublic networks. *Information, Communication & Society, 19*(3), 397–418.

Jackson, S. J., Bailey, M., & Foucault-Welles, B. (2020). *#Hashtag activism: Networks of race and gender justice*. MIT Press.

Jenkins, H. (2006). *Convergence culture: Where old and new media collide*. New York University Press.

Jesus Film Project. (2020). Jesus Film Project. [Webpage]. https://www.jesusfilm.org/

Jore, T. (2013). *The Christian commons: Ending the spiritual famine of the global church*. Distant Shores Media.

Jore, T. (2017). Personal interview. Google Hangout.

Joyce, K. (2013).*The child catchers: Rescue, trafficking, and the new gospel of adoption*. Public Affairs.

Kenny, C., & Sandefur, J. (2013). Can Silicon Valley save the world? *Foreign Policy*, (201), 72–77.

Kent, K. G. (2017, April 28). The "crisis" in the blogosphere. [Blog post]. https://revolfaith.com/2017/04/27/whos-in-charge-of-monks-nailing-theses-to-church-doors/

Kessler, S. (2018). *Gigged: The gig economy, the end of the job and the future of work*. St. Martin's Press.

Kilde, J. H. (2002). *When church became theater: The transformation of evangelical architecture and worship in nineteenth-century America*. Oxford University Press.

Kilde, J. H. (2006). Reading megachurches: Investigating the religious and cultural work of church architecture. In L. P. Nelson (Ed.), *American sanctuary: Understanding sacred spaces* (pp. 225–245). Indiana University Press.

King Jr., M. L. (1963/1994). *Letter from the Birmingham jail*. Harper San Francisco.

Kisner, J. (2013). Jesus raves. *N + 1*. https://nplusonemag.com/issue18/essays/jesusraves/

Kipling, R. (1899). The white man's burden. *McClure's Magazine*.

Klassen, P., & Lofton, K. (2013). Material witnesses: Women and the mediation of Christianity. In M. Lovheim (Ed.), *Media, religion, and gender: Key issues and new challenges* (pp. 52–65). Routledge.

Kristian, B. (2020, May 21). Is QAnon the newest American religion? *Newsweek*. https://theweek.com/articles/915522/qanon-newest-american-religion

Kruse, K. (2005). *White flight: Atlanta and the making of modern conservatism*. Princeton University Press.

Kruse, K. (2015). *One nation under God: How corporate America invented Christian America*. Basic Books.

LaBrant, C., & LaBrant, S. (2018). *Cole + Sav: Our surprising love story*. Thomas Nelson.

The LaBrant Fam. (2017, July 15). Our wedding video!!! Vows to 4 year old daughter [Video]. YouTube. https://www.youtube.com/watch?v=MK5zVLYXzyg

LaBrant, S. [sav.labrant]. (2018, June 17). I couldn't be more obsessed with these two. She loves him so incredibly much and has him wrapped around her [Instagram post]. Instgram. https://www.instagram.com/p/BkJE-1qAM71/?utm_medium=share_sheet

Laughlin, C. (2020). #AmplifyWomen: The emergence of an evangelical feminist public on social media. *Feminist Media Studies*.

LeCrae. (2016, October 26). The pains of humanity have been draining me. *The Huffington Post*. https://www.huffpost.com/entry/i-declare-black-lives-matter_b_5808be36e4b0dd54ce385412

Lee, M. (2018, December 13). Max Lucado reveals past sexual abuse at evangelical #MeToo summit. *Christianity Today*. https://www.christianitytoday.com/news/2018/december/metoo-evangelicals-abuse-beth-moore-caine-lucado-gc2-summit.html

Lenae, M. (Host). (2016–present). *Christ over Culture* [Audio podcast].

Lenae, M. (2020). Personal interview. Zoom.

Life.Church (2012). *LifeChurch.tv's Vision and Values* [Video]. YouTube. https://www.youtube.com/watch?v=YwbPnLz5WOY

Life.Church. (2018). Life.Church [homepage]. http://www.life.church

LifeChurch.tv (2014). Equipping churches. [webpage] https://www.lifechurch.tv/churches/

Lincoln, C., & Mamiya, L. (1990). *The Black church in the African American experience*. Duke University Press.

Luhrmann, T. (2012). *When God talks back: Understanding the American evangelical relationship with God*. Vintage Books.

Lorenz, T. (In press). *Extremely online: Gen Z, the rise of influencers, and the creation of a new American dream*. Simon & Schuster.

MacGillis, A. (2021). *Fulfillment: Winning and losing in one-click America*. Farrar, Straus, and Giroux.

MacKenzie, D. A., & Wajcman, J. (1999). *The social shaping of technology* (2nd ed.). Open University Press.

Manning, C. (1999). *God gave us the Right: Conservative Catholic, Evangelical Protestant, and Orthodox Jewish women grapple with feminism*. Rutgers University Press.

Maples, J. (2020, June 27). Now we know why Jen Hatmaker is gay-affirming. *Reformation Charlotte*. https://reformationcharlotte.org/2020/06/27/now-we -know-why-jen-hatmaker-is-gay-affirming-her-daughter-is-a-lesbian/

Marcus, G. (1996). Ethnography in/of the world system: The emergence of a multi-sited ethnography. *Annual Review of Anthropology* (24), 95–117.

Marsden, G. (2006). *Fundamentalism and American culture*. Oxford University Press.

Marti, G. (2005). *A mosaic of believers: Diversity and innovation in a multiethnic church*. Indiana University Press.

Marti, G. (2020). *American blindspot: Race, class, religion, and the Trump presidency*. Rowman & Littlefield.

Martin, A. (2017). Personal interview. Sunnyvale, CA.

Marwick, A. (2013). *Status update: Celebrity, publicity, and branding in the social media age*. Yale University Press.

Marwick, A. (2015). Instafame: Luxury selfies in the attention economy. *Public Culture*, 27(1), 137–160.

Marwick, A., & Boyd, D. (2012). I tweet honestly, I tweet passionately: Twitter users, context collapse and the imagined audience. *New Media & Society* 13(1), 114–133.

McAlister, M. (2018). *The Kingdom of God has no borders: A global history of American evangelicals*. Oxford University Press.

McCarver, V. (2012). The new oxymoron: Socially conservative feminism. *Women & Language*, 35(1), 57–76.

McConnell, S. (2009). *Multi-site churches: Guidance for the movement's next generation*. B&H Publishing Group.

McCorvey, J. (2015). Special report: Black in Silicon Valley. *Fast Company*. December/January.

McGee, P. L. (2017). *Brand® new theology : The Wal-Martization of T. D. Jakes and the New Black Church*. ORBIS.

McManus, E. (2006). Foreword. In G. Surratt, G. Ligon, & W. Bir (Eds.) *The multi-site church revolution: Being one church in many locations*. Zondervan.

McNeil, J. (2020). *Lurking: How a person became a user*. Farrar, Straus and Giroux.

McRobbie, A. (2009). *The aftermath of feminism: Gender, culture and social change*. Sage Publications.

Mencken, H. L. (1925, June 29). Homo Neadrethalis. *The Baltimore Sun*.

Menzie, N. (2017, April 24). Black Christian women break silence on facing sexism and racism in church. *Faithfully Magazine*. http://faithfullymagazine.com/black-christian-women-break-silence-facing-sexism-racism-church/

Merchant, B. (2017). *The one device: The secret history of the iPhone*. Little Brown.

Meserko, V. M. (2015). The pursuit of authenticity on Marc Maron's WTF podcast. *Continuum: Journal of Media & Cultural Studies, 29*(6), 796–810.

Milano, A. [@Alyssa_Milano]. (2017, October 15). If you've been sexually harassed or assaulted write "me too" as a reply to this tweet [Tweet]. Twitter.com. https://twitter.com/alyssa_milano/status/919659438700670976?lang=en

Miller, D., & Slater, D. (2001). *The Internet: An ethnographic approach*. Routledge.

Mobile Ministry Forum. (2017). Mobilize for the unreached. http://mobileministryforum.org/

Mohler, A. (2018, May 23). The wrath of God poured out. https://albertmohler.com/2018/05/23/wrath-god-poured-humiliation-southern-baptist-convention

Monroe, R. (2017). #Vanlife, the bohemian social media movement. *The New Yorker*. https://www.newyorker.com/magazine/2017/04/24/vanlife-the-bohemian-social-media-movement

Moore, B. (2018a). About Beth Moore. *Living Proof Ministries* [Website]. http://www.lproof.org/about

Moore, B. (2018b, May 3). A letter to my brothers. [Blog post]. The LPM Blog. https://blog.lproof.org/2018/05/a-letter-to-my-brothers.html

Moore, B. [@BethMooreLPM]. (2019, February 10). We understand how you feel. We didn't want to know about sexual abuse either [Tweet]. Twitter.com. https://twitter.com/bethmoorelpm/status/1094720702190678016?lang=en

Morrison, L. (2019). *Be the bridge: Pursuing God's heart for racial reconciliation*. WaterBrook.

Moufe, C. (1999). Deliberative democracy or agonistic pluralism?. *Social research*, 745–758.

Murrell, S. (2011). *Wikichurch: Making discipleship, engaging, empowering, and viral*. Charisma House.

Nelsen, H., & Nelsen, A. (1975). *Black church in the sixties*. University Press of Kentucky.

Nichols, B. (1991). *Representing reality: Issues and concepts in documentary*. Indiana University Press.

Noll, M. (1994). *The scandal of the evangelical mind*. Eerdmans.

Noll, M. (2001). *American evangelical Christianity: An introduction*. Blackwell Publishers.

Nord, D. (2004). *Faith in reading: Religious publishing and the birth of mass media in America*. Oxford University Press.

Northland (2011). What is worship? [Video]. http://www.northlandchurch.net /articles/how_why_and_where_we_worship/

O'Leary, A. (2013, July 26). In the beginning was the Word; now the Word is on an app. *New York Times*.

Onishi, B. (2019, April 19). The rise of #Exvangelical. *Religion & Politics*. https:// religionandpolitics.org/2019/04/09/the-rise-of-exvangelical/.

Park, J. Z., & Davidson, J. C. (2020). Decentering whiteness in survey research on American religion. In G. Yukich & P. Edgell (Eds.), *Religion is raced: Understanding American religion in the twenty-first century* (pp. 251–274). New York University Press.

Peck, R. (director). (2016). *I am not your negro* [Documentary film]. Velvet Film.

Perry, S. (2013). She works hard(er) for the money: Gender, fundraising, and employment in evangelical parachurch organizations. *Sociology of Religion, 74*(3), 392–415.

Perry, J. H. (2018). *Gay girl, good God: The story of who I am and who God has always been*. B & H Publishing Group.

Pew Research Center. (2015, May 12). *America's Changing Religious Landscape*. https://www.pewforum.org/2015/05/12/americas-changing-religious-landsc ape/

Phillips, W. (2015). *This is why we can't have nice things: Mapping the relationship between online trolling and mainstream culture*. MIT Press.

Pimblott, K. (2016). Black power and black theology in Cairo, Illinois. In C. D. Cantwell, H. W. Carter, & J.

G. Drake (Eds.), *The pew and the picket line: Christianity and the American working class* (pp. 115–142). University of Illinois Press.

Pinn, A. B. (2002). *The Black Church in the civil rights era*. Orbis Books.

Poell, T., & van Dijck, J. (2018). Social media and new protest movements. In J. Burgess, A. Marwick, & T. Poell (Eds.), *The sage handbook of social media* (pp. 546–561). Sage.

Posner, S. (2020). *Unholy: Why white evangelicals worship at the altar of Donald Trump*. Random House.

Pratt, Z. (2016). Here's what we mean by unreached people and places. *The*

International Missions Board. https://www.imb.org/2016/11/22/what-do-we
-mean-by-unreached-peoples-and-places/

Propel Women. (2021). Propel Women [website]. https://www.propelwomen.org
/index.php

Pulos, J. (2017). Personal interview. Skype.

Raymond, E. S. (2001). *The cathedral and the bazaar: Musings on Linux and
open source by an accidental revolutionary.* O'reilly Media.

Relevant. (2019, October 21). John MacArthur told Beth Moore to "go home."
Relevant Magazine. https://relevantmagazine.com/current16/john-macarth
ur-told-beth-moore-to-go-home-for-having-the-audacity-to-preach-the-gos
pel-and-help-people/

Rheingold, H. (1993). *Virtual community: Homesteading on the electronic
frontier.* Addison-Wesley Publishing Company.

Rice, J. (2009). *The church of Facebook: How the hyperconnected are redefining
community.* David C. Cook.

Richardson, A. V. (2020). *Bearing witness while Black: African Americans,
smartphones & the new protest #journalism.* Oxford University Press.

Robertson, C. (2018, March 9). A quiet exodus: Why Black worshipers are leaving white evangelical churches. *New York Times.* https://www.nytimes.com
/2018/03/09/us/blacks-evangelical-churches.html

Rodino-Colocino, M. (2012). Man up, woman down: Mama grizzlies and antifeminist feminism during the year of the (conservative) woman and beyond.
Women & Language, 35(1), 79–95.

Rogers, E. M., & Larsen, J. K. (1985). *Silicon Valley fever: Growth of high technology culture.* Allen and Unwin.

Roof, W. C. (1993). *A generation of seekers: The spiritual journeys of the Baby
Boom generation.* HarperSanFrancisco.

Roof, W. C. (1999). *Spiritual marketplace: Baby Boomers and the remaking of
American religion.* Princeton University Press.

Rubin, J. (2017). *Perishing heathens: Stories of Protestant missionaries and
Christian Indians in Antebellum America.* University of Nebraska Press.

Salvatore, N. (2005). *Singing in a strange land: C. L. Franklin, the Black church
and the transformation of America.* Little Brown.

Sanneh, L. (2008). *Disciples of all nations: Pillars of world Christianity.* Oxford
University Press.

Schulte, S. R. (2013). *Cached: Decoding the internet in global popular culture.*
New York University Press.

Schultze, S. R. (1991). *Televangelism and American culture: The business of
popular religion.* Baker Book House.

Shellnutt, K. (2014, September 29). Christine Caine, Liberty University to launch
"lean in"-type program for Christian women. *Christianity Today.* https://

www.christianitytoday.com/ct/2014/september-web-only/christine-caine-libe
rty-university-to-launch-lean-in-type-p.html

Smietana, B. (2014). Who owns the sermons? *Christianity Today, 58*(1), 48–51.

Smietana, B. (2018, March 22). Bill Hybels accused of sexual misconduct by
former Willow Creek leaders. *Christianity Today.* https://www.christianityto
day.com/news/2018/march/bill-hybels-misconduct-willow-creek-john-nancy
-ortberg.html

Smietana, B. (2021, March 9). Bible teacher Beth Moore, splitting with Lifeway,
says, "I am no longer a Southern Baptist." https://religionnews.com/2021/03
/09/bible-teacher-beth-moore-ends-partnership-with-lifeway-i-am-no-longer
-a-southern-baptist/

Southern Baptist Convention (2012). Resolution on same sex marriage. http://
www.sbc.net/resolutions/1224/on-samesex-marriage-and-civil-rights-rhetoric

Spinelli, M., & Dann, L. (2019). *Podcasting: The audio media revolution.*
Bloomsbury Academic.

Squires, C. R. (2002). Rethinking the Black public sphere: An alternative vocab-
ulary for multiple public spheres. *Communication Theory, 12*(4), 446–468.

Stallman, R. (2010). *Free software free society: Selected essays of Richard Stallman* (2nd ed.). Free Software Foundation.

Stephenson, M. (2011). *Web-empowered ministry: Connecting with people
through websites, social media, and more.* Abingdon Press.

Stetzer, E. (2018, November 28). Slain missionary John Chau prepared much
more than we thought, but are missionaries still fools? *Washington Post.*
https://www.washingtonpost.com/religion/2018/11/28/slain-missionary-john
-chau-prepared-much-more-than-we-thought-his-case-is-still-quandary-us
-missionaries/

Stewart, J. B. (2013, March 15). Looking for a lesson in Google's perks. *New York
Times.* https://www.nytimes.com/2013/03/16/business/at-google-a-place-to
-work-and-play.html

Streeter, T. (2005). The moment of wired. *Critical Inquiry, 31*(4), 755–779.

Streeter, T. (2011). *The net effect: Romanticism, capitalism, and the internet.* New
York University Press.

Sterne, J. (2003). Bourdieu, technique and technology. *Cultural Studies, 17*(3/4),
367.

Sunstein, C. (2017). *#Republic: Divided democracy in the age of social media.*
Princeton University Press.

Surratt, G., Ligon, G., & Bird, W. (2006). *The multi-site church revolution: Being
one church in many locations.* Zondervan.

Sweet, L. (2012). *Viral: How social networking is poised to ignite revival.*
WaterBrook Press.

Sweetman, D. (2017). Personal interview. Los Angeles, CA.

Takhteyev, Y. (2012). *Coding places: Software practice in a South American city.* MIT Press.

Thurman, H. (1996). *Jesus and the disinherited.* Beacon Press.

Tisby, J. (2019). *The color of compromise: The truth about the American church's complicity in racism.* Zondervan Reflective.

Tisby, J. (2020). Personal interview. Zoom.

Tisby, J., & Burns, T. (Hosts). (2020a, June 1). George Floyd and an unjust nation on the brink [Podcast episode]. *Pass the Mic* [Podcast]. The Witness Foundation.

Tisby, J., & Burns, T. (Hosts). (2020b, June 22). We are not shields or stools [Podcast episode]. *Pass the Mic* [Podcast]. The Witness Foundation.

Tisby, J., & Burns, T. (Hosts). (2020c, July 20). #LeaveLoud with Ally Henny and Zee Johnson [Podcast episode]. *Pass the Mic* [Podcast]. The Witness Foundation.

Tisby, J., & Burns, T. (Hosts). (2020d, July 27). Black Christians, don't negotiate with your dignity [Podcast episode]. *Pass the Mic* [Podcast]. The Witness Foundation.

Tsing, A. (2012). On nonscalability: The living world is not amenable to precision-nested scales. *Common Knowledge, 18*(3), 505–524.

Tufekci, Z. (2017). *Twitter and tear gas: The power and fragility of networked protest.* Yale University Press.

Turner, F. (2006). *From counterculture to cyberculture: Stewart Brand, the Whole Earth network, and the rise of digital utopianism.* University of Chicago Press.

Voskamp, A. (2017, May 5). About who's in charge of the Christian blogosphere: An ongoing conversation. [Blog post]. AnnVoskamp.com [Website]. https://annvoskamp.com/2017/05/about-whos-in-charge-of-the-blogosphere-an-ongoing-conversation/

Wachter-Boettcher, S. (2017). *Technically wrong: Sexist apps, biased algorithms, and other threats of toxic tech.* W. W. Norton and Company, Inc.

Walker-Barnes, C. (2019). *I bring the voices of my people: A womanist vision for racial reconciliation.* Eerdmans.

Walton, J. (2009). *Watch this! The ethics and aesthetics of black televangelism.* New York University Press.

Wark, M. (2004). *A hacker manifesto.* Harvard University Press.

Warner, M. (2002). *Publics and counterpublics.* Zone Books.

Warren, R. (1995). *The purpose driven church: Growth without compromising your mission and message.* Zondervan Publishing.

Warren. R. [@RickWarren]. (2014, December 5). Racism is a SIN problem, not a SKIN problem. My thoughts http://on.fb.me/1zhFc7F on Facebook [Tweet]. https://twitter.com/rickwarren/status/540953924367163393?lang=en

Warren, T. H. (2017, April 27). Who is in charge of the Christian blogosphere?

Christianity Today. http://www.christianitytoday.com/women/2017/april/wh os-in-charge-of-christian-blogosphere.html

Warner, M. (2002). *Publics and counterpublics*. Zone Books.

Wasik, B. (2015, June 5). Welcome to the age of digital imperialism. *New York Times Magazine*. http://www.nytimes.com/2015/06/07/magazine/welcome -to-the-age-of-digital-imperialism.html

Weber, M. (1968). *Economy and society*. University of California Press.

Whitehead, D. (2015). The evidence of things unseen: Authenticity and fraud in the Christian mommy blogosphere. *Journal of the American Academy of Religion, 83*(1), 120–150.

Williams, K. (2017). Personal interview. Skype.

Winner, L. (1986). *The whale and the reactor: A search for limits in an age of high technology*. University of Chicago Press.

Wise Rowe, S. (2020). *Healing racial trauma: The road to resilience*. InterVarsity Press.

Wright, A. (2017). Personal interview. Skype.

Wuthnow, R. (1998). *After heaven: Spirituality in American since the 1950s*. University of California Press.

Yorke, J. (2017). Personal interview. Skype.

Young, M. P. (2006). *Bearing witness against sin: The evangelical birth of the American social movement*. University of Chicago Press.

YouVersion. (n.d.). Retrieved February 2020. https://www.youversion.com/the -bible-app/

Index

Moore, Beth, 119, 122–129
Moore, Russell, 13, 98, 115–116, 137–138
multisite churches, 28–31, 42–43, 45

Nashville statement, 115
National Association of Evangelicals (NAE), 7, 171n3
Negroponte, Nicholas, 81, 83. *See also* One Laptop per Child Program
neo-evangelical, 6–7
NeverTrump movement, 113, 119, 178n10
Northland Church, 27, 29–30

One Laptop per Child Program, 81, 83. *See also* Negroponte, Nicholas
online church, 31–34, 37, 43–49
Osteen, Joel, 27, 141, 160

Palin, Sarah, 4, 137, 178n15
parasocial interaction, 101–102
Pass the Mic podcast, 130, 141, 144–145, 149, 153–154, 162
Patterson, Paige, 127
Perry, Jackie Hill, 114–115
Piper, John, 116
podcasting, 139–145, 160, 162
pornography, 69
Promise Keepers, 133, 135–136, 151, 156

QAnon, 171–174

racial reconciliation movement, 131, 135–139, 145, 150–151, 156–157
racism: in evangelical culture, 70–71, 125–126, 131, 137–139, 142, 145; and intersectionality, 159–160, 179n19; in missions work, 77–80; and racial trauma in evangelical churches, 148–154; role of in the creation of the Black Church 132–135; in the technology industry 70–71. *See also* racial reconciliation movement
redemptive entrepreneurship, 65–72
Rheingold, Howard, 86–87, 97
Romney, Mitt, 160

Saddleback Church, 24–26, 28–29, 46. *See also* Warren, Rick
Scopes Trial, 5, 6
sexism: in evangelical culture, 103–104, 121, 124–125, 158–160; in the technology industry, 55
Silicon Valley: bias in, 70–71, 95–96, 172n13; Christian critique of, 68; cultural influence of, 13, 34, 49, 51–52, 57–65, 81; history of, 52–56
sola scriptura, 81
Southern Baptist Convention (SBC), 98, 113, 123–128, 136–138, 160
Stetzer, Ed, 75

televangelism, 8, 31, 122, 135, 138, 171n5
transphobia, 115
trolling, 44, 142
Trump, Donald: evangelical support of, 5, 113, 124, 128, 137–138; and QAnon, 163; racial politics of, 25, 147, 152, 157
Truth's Table podcast, 158–160
Twitter: and digital habitus, 9, 14; history of, 54; relationship of to online church, 44; role in creating activist spaces, 99, 117–122; role in the #MeToo movement, 125–126

Uwan, Ekemini, 125, 158, 160

virtual reality (VR), 20, 48, 52, 65
Voskamp, Ann, 121

Warren, Kay, 123
Warren, Rick, 24–27, 29, 70, 137, 140, 141
Warren, Tish Harrison, 118–120
Web 2.0, 54
Westboro Baptist Church, 3
White, Paula, 138
"The White Man's Burden," 76–77
Willow Creek Community Church, 126, 171n4. *See also* Hybels, Bill
Wired Magazine, 53

YouVersion Bible App, 21, 36, 38, 41–45

Founded in 1893,
UNIVERSITY OF CALIFORNIA PRESS
publishes bold, progressive books and journals
on topics in the arts, humanities, social sciences,
and natural sciences—with a focus on social
justice issues—that inspire thought and action
among readers worldwide.

The UC PRESS FOUNDATION
raises funds to uphold the press's vital role
as an independent, nonprofit publisher, and
receives philanthropic support from a wide
range of individuals and institutions—and from
committed readers like you. To learn more, visit
ucpress.edu/supportus.

9 780520 379671